CITRUS DISEASES AND DISORDERS

CITRUS DISEASES AND DISORDERS

An Alphabetized Compendium with
Particular Reference to Florida

L. C. Knorr

Scab

Melanose

Russet

Oleocellosis

Greasy Spot Rind Blotch

A University of Florida Book

The University Presses of Florida
Gainesville • 1973

Library of Congress Cataloging in Publication Data

Knorr, Louis Carl, 1914-
　Citrus diseases and disorders.

　"A University of Florida book."
　Bibliography: p.
　1.　Citrus fruits—Diseases and pests—Dictionaries.
2. Citrus fruits—Diseases and pests—Florida—
Dictionaries. I. Title.
SB608.C5K63　　　634'.3　　　73-17462
ISBN 0-8130-0383-0

PRINTED IN FLORIDA, U.S.A.

In dedication, and as partial restitution, to Lila
G. Knorr and our children Carol Jean and Guthrie

CONTENTS

PREFACE

The pathologist encounters difficulties in precisely defining *disease*. Yet the agriculturist has little trouble in recognizing an ailing plant. To him, disease is any nonentomological abnormality that diminishes or destroys the economic value of his crop. It is this practical and widely inclusive sense that has been adopted in the text.

It is this same sense of practicality that governs the discussion. The author's allegiance is to the grower; consequently, only that part of pathological research is included that seeks to maximize yields and pack-outs or to minimize costs. For that reason, this offering may be of interest to all who are devoted to the crop: growers, agricultural advisors, students, and specialists—particularly if they do not have the library resources that exist at only the few large citrus research institutions of the world.

In general, the length of each subject is proportional to its economic importance. Exceptions occur in the case of recently discovered diseases, the consequences of which remain unknown.

No treatments are intended of problems relating to entomology and nutrition, though in a few instances (e.g., leprosis and yellow spot) such problems have been included, especially if at one time they were considered to involve pathogens.

Only those diseases and disorders are discussed that have been reported from Florida. The majority of maladies, however, occur throughout the world where citrus is grown. Those that are presently not known in Florida are presented in Appendix 1. With respect to the major exotic diseases, the reader is referred to the author's "Serious Diseases of Citrus Foreign to Florida" (Bulletin 5, Florida State Department of Agriculture, Division of Plant Industry, Gainesville, 1965).

The arrangement of the subjects is alphabetical according to the common names of diseases and disorders. Interspersed in the same order are the scientific names of disease-producing organisms; thus is supplied a checklist of all pathogens recorded on Florida citrus.

The intent has been to write in the language of the grower; however, the needs of precision have led unavoidably to the use of terms that may be unfamiliar. For the explanation of scientific words and concepts, as well as of various colloquialisms, a glossary is appended.

Warning on the use of pesticides

Control recommendations involving fungicides, pesticides, and nematocides are based on information current during the writing of the text. To keep abreast of ever changing legal restrictions, the grower is advised to consult with qualified agricultural specialists or consultants in designing his yearly spray program. Additional assistance is provided by the annually revised "Florida Citrus Spray and Dust Schedule" made available by the Florida Department of Citrus, Lakeland, Fla. At all times the labels of agricultural chemicals should be followed explicitly for the prevention of harm to pesticide handlers, for the restriction of pesticide drift to other crops and properties, and for the avoidance of illegal pesticide residues.

Acknowledgments

All chapters have been reviewed, checked for accuracy, and brought up to date by appropriate specialists. The author is indebted for assistance received from the following collaborators: Dr. C. A. Anderson, Dr. R. F. Brooks, Dr. G. E. Brown, Dr. E. C. Calavan, Dr. D. V. Calvert, Dr. H. D. Chapman, Dr. J. F. L. Childs, Dr. M. Cohen, Dr. E. P. DuCharme, Mr. R. P. Esser, Dr. A. W. Feldman, Dr. H. W. Ford, Dr. S. M. Garnsey, Dr. W. Grierson,

Dr. C. I. Hannon, Dr. R. B. Johnson, Dr. L. J. Klotz, Dr. R. C. J. Koo, Dr. C. D. Leonard, Mr. A. A. Mc-Cornack, Dr. M. H. Muma, Dr. J. H. O'Bannon, Mr. C. Poucher, Dr. P. C. Reece, Dr. R. L. Reese, Dr. H. J. Reitz, Dr. A. S. Rhoads, Dr. R. E. Schwarz, Dr. W. A. Simanton, Dr. I. Stewart, Dr. R. F. Suit, Dr. A. C. Tarjan, Dr. A. L. Taylor, Dr. J. M. Wallace, Dr. I. W. Wander, and Dr. J. O. Whiteside.

Indebtednesses to contributors of photographs are acknowledged in the figure captions. Those figures without credit lines are mainly the work of Harriet Long and Madge Collier.

Gratefully acknowledged is the considerable editorial assistance of Dr. J. O. Whiteside.

L. C. Knorr

May 17, 1971
Agricultural Research and Education Center
Institute of Food and Agricultural Sciences
University of Florida
Lake Alfred, Florida

Postscript

The author is grateful to the University of Florida Press for having rescued the manuscript from uncertainties of funding through regular channels. During the ensuing delay in publication, opportunity was taken to update the text by the addition of footnotes reflecting recent advances in knowledge. Two more appendices were also added. Appendix 2 lists the common names of citrus varieties referred to in the text and gives their scientific equivalents according to the two widely used systems of Tanaka and Swingle. Appendix 3 is an index and a guide to parts of the tree affected by the various diseases and disorders; its purpose is to give assistance in diagnosing diseases in the field and to direct the user to pertinent discussions in the text.

March 31, 1973
United Nations Development Programme
GPO Box 618
Bangkok 5, Thailand

CITRUS DISEASES AND DISORDERS

AEGERITA. Genus of the deuteromycetous fungi. *A. webberi* Fawc. is parasitic on whitefly larvae. See **Entomogenous Fungi.**

AGING OF FRUIT. See **Stem–End Rind Breakdown.**

ALGAL DISEASE. An algal-induced bark splitting, especially in trees of lemon and lime; also a rind spotting on senescent fruits of miscellaneous varieties. *SYN.*—green scurf, red rust.

Classically, *Cephaleuros virescens* is an example of a pathogen that is not a parasite; economically, it is the cause of a disease that may limit the growing of lemons and limes.

Plants affected

The causal organism may be found on trees of all varieties, but it is destructive mainly to trees of lemon, lime, and certain of the specialty fruits (188). Occasionally in the counties of Lee, Sarasota, Manatee, Pinellas, and Hillsborough, it may be destructive to all varieties of citrus. In Florida, *C. virescens* is known to occur on some 50 hosts in addition to *Citrus* spp. On some plants it is merely a harmless epiphyte (279).

Symptoms

In contrast to most other pathogenic thallophytes, the causal organism of algal disease is conspicuous in advance of symptoms. Colonies at time of fruiting in summer are brick-red, raised, velvety, and often doughnut-shaped; at other times of the year, they are grayish-green. Individual colonies have an average diameter of ½ in. but on coalescing with other colonies, they may cover branches in a continuous sheath.

The first symptom on branches (the site of principal damage) is a thickening of the bark around the colonies. In time, the raised bark cracks and produces small irregularly shaped platelets or shreds (Fig. 1A). Terminal growth of branches is restricted, causing leaves to become chlorotic and to drop. When conditions are favorable for the rapid development of algal disease, branches up to 2 in. in diameter may be killed.

Colonies produce little damage on leaves (Fig. 2). The velvety patches occur on either surface. At times colonies wither and are blown away, uncovering small depressions on the leaf surface.

Occasionally, fruits of orange, grapefruit, lemon, and lime may be spotted with blackish, circular, 1/16–1/4-in.-wide colonies of the same organism (176, 368). Under the hand lens, colonies resemble clumps of a miniature branching moss (Fig. 1B, C). Other than producing blemishes, which can be removed by brushing, these growths are of no consequence, especially since they occur mainly on overripe, unmarketable fruits.

Cause

Cephaleuros virescens Kunze, the cause of algal disease, belongs to that group of green algae that are characterized by living in air and possessing hematochrome (166). *C. virescens* is synonymous with *Mycoidea parasitica* Cunningham and *Cephaleuros mycoidea* Karsten, and probably also with *Phyllactidium tropicum* Mobius and *Cephaleuros parasiticus* Karsten.

In Florida, the alga occurs independently, but elsewhere in the world it may be associated with a fungus to form the lichen *Strigula complanata*.

On leaves, penetration by rain-borne zoospores appears to take place through wounds. On germination, zoospores form a subcuticular or subepidermal thallus which sends up sterile and sporangia-bearing hairs above the surface of the leaf, giving the alga its velvety appearance. Epidermal cells beneath the algal patch develop a brown discoloration. The cycle from zoospore to sporan-

1

gium-bearing thallus, as studied in *Magnolia,* takes about 8 months; presumably the duration in citrus is the same.

Citrus leaves unfolding in spring apparently remain free of infection until the end of summer. In late September or early October, infections become evident on these 4–5-month-old leaves. Two months later, prominent thalli beset with sterile hairs are conspicuous. Thalli continue to enlarge throughout the winter, and by the end of May—approximately 8 months following infection—both stalked and sessile sporangia are present and begin to emit biciliate zoospores. Thalli continue to grow and release successive crops of sporangia for as long as leaves remain on the tree and for some time after leaves have dropped (373).

The mode of infection in bark has not been studied, but it is probable that zoospores lodge in lenticels and wounds. Since the alga is capable of manufacturing its own food, no true parasitism is involved. Damage to the host results apparently from mechanical rupturing of the bark due to pressure generated by the multiplying algal cells.

Control

If trees receive routine copper sprays as applied for the control of melanose, scab, and greasy spot, algal disease does not become a problem. Thor-

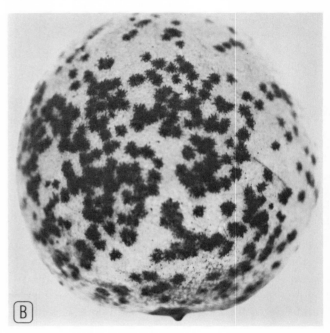

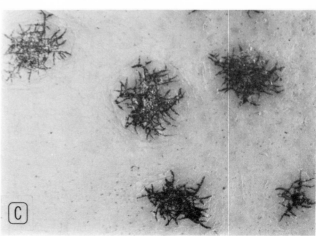

FIG. 1. *ALGAL DISEASE.* A. Left: Branch of Meyer lemon showing felty patches of *Cephaleuros virescens.* Center: Cracking of the bark following invasion. Right: Ultimate shredding of the bark. B. Algal spot on senescent orange. C. Magnification of algal colony on rind of lemon.

ough spraying of the wood is, of course, a requisite for control.

In groves where algal disease has gotten out of hand, a clean-up program is needed to bring the disease under control. A neutral copper spray (1 lb. metallic Cu/100 gal. water), plus oil or a sticker-spreader to wet the colonies, has been found effective if applied at the late dormant stage (sometime in February), again at the postbloom

stage, and finally at the stage when the alga turns red (approximately in June) (318). Once the clean-up program is successful, the disease can be kept in check routinely with annual postbloom applications of copper.

ALTERNARIA. Genus of the deuteromycetous fungi. *A. citri* Ellis & Pierce (27) is associated with **Alternaria Leafspot** and **Black Rot** (which see).

ALTERNARIA LEAFSPOT. An important foliage disease in the nursery, particularly on rough lemon and Rangpur lime seedlings. *SYN.*—anthracnose (an improper former usage).

Certain foliage diseases need to be controlled in the nursery if seedlings intended for rootstocks are to grow straight and rapidly. The most important of these diseases is Alternaria leafspot (Fig. 3), more commonly (but erroneously) referred to in Florida as anthracnose. This spot is commercially important only on rough lemon and Rangpur lime. Lesions develop following infection of young leaves and appear within the leaf blade or along the margins. Individual spots seldom exceed 1 in. in diameter, but the confluence of several spots may involve much of the leaf area. Color of the necrotic tissue ranges from light to dark brown except at the edges, where the color is darker than at the centers. The shape is round to somewhat angular. When a spot traverses a major vein, the lesion tends to extend out along the vein. The spot is usually surrounded by a chlorotic halo. Severe attacks lead to leaf drop and to excessive branching that produces an undesirable bushy plant.

With age, lesions become speckled with black dots, which are the fruiting stalks of the Alternaria fungus along with the fruiting structures of other fungi, principally *Colletotrichum gloeosporioides.* These dots are often arranged in concentric rings.

The cause of Alternaria leafspot was at one time attributed to *C. gloeosporioides,* but later studies demonstrated that the disease is caused by *Alternaria citri* Ellis & Pierce (278). Both fungi can cause infections in citrus that remain invisible until tissues are weakened or killed by other agents (see **Symptomless Infections**).

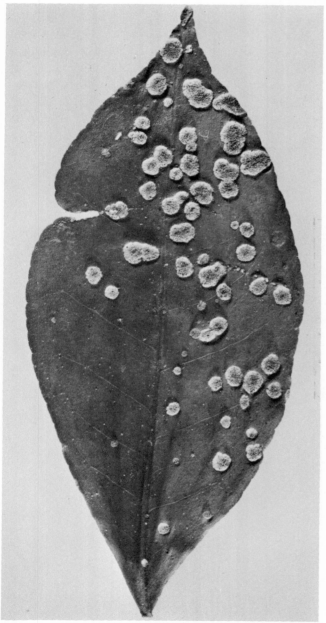

FIG. 2. *ALGAL DISEASE.* Colonies of *Cephaleuros virescens* on leaf of *Aeglopsis chevalieri.*

Typical spots have been produced in the greenhouse by placing drops of a spore suspension of *A. citri* on young leaves. In 4 days, leaves develop small, pale green areas with brownish centers and in 4 weeks reproduce the spotting seen in the field. When inoculum consisted of *A. citri* from fruits affected by **Black Rot** (which see), no spots developed on leaves, thus suggesting the existence of different strains of *A. citri*.

The current practice for controlling Alternaria leafspot, as well as for controlling scab, is to spray rough lemon and Rangpur lime seedlings every 2 weeks with neutral copper (¾ lb. metallic content/100 gal. water). Control has also been reported from weekly applications of Captan (6 lb./100 gal. water) and from tri-weekly applications of zineb (2 lb./100 gal. water) (104, 155).

ALTERNARIA ROT. See **Black Rot.**

ALTERNARIA SPOT. See **Black Rot.**

AMMONIATION. See **Exanthema.**

ANTHRACNOSE. A twig dieback and a fruit blemish long attributed (erroneously so, in the light of present knowledge) to a fungus. *SYN.*—wither tip, anthracnose stain, anthracnose spot.

The meaning of anthracnose has changed over the years. At one time, the name was applied to troubles thought to be initiated by the fungus *Colletotrichum gloeosporioides* Penz. (perfect stage= *Glomerella cingulata* [Stonem.] Spauld. & von Schrenk). Although the term is still used, it is inapt

FIG. 3. *ALTERNARIA LEAFSPOT.* A leafspotting disease of rough lemon; formerly called anthracnose.

because it is applied to three separate problems: a virulent disease of Key limes caused by *Gloeosporium limetticola* (see **Lime Anthracnose**); a foliar disease of rough lemon and Rangpur lime (see **Alternaria Leafspot**); and a dieback and fruit blemish in which *C. gloeosporioides* is present either as a latent infection or as a secondary invader. It is the third condition that will be discussed here.

Wither tip was originally described as a gradual yellowing, parching, and dropping of leaves followed by a dying back of twigs and branches. At times, leaves wilted and dried out rapidly and remained hanging on the tree, giving portions of the canopy the appearance of having been scorched by fire. An examination of affected twigs showed a slight gumming and a sharp line of separation between healthy and killed tissues. Dead terminals were speckled with many minute black pustules—the acervuli or fruiting bodies of the fungus.

Isolations from diseased tissues usually yielded the anthracnose fungus, and it was assumed that *C. gloeosporioides* was the cause. Today, however, this fungus is no longer regarded as an initiator of disease. *C. gloeosporioides* is presently considered either as a secondary invader of weakened tissue or as a nonpathogenic or latent infection that becomes evident only after the fungus has sporulated in dead tissue (1).

Wither tip is no longer the problem it was once considered to be. The change in status has come about through the recognition that the fungus is nonpathogenic and that symptoms are the effects of other agencies. Most commonly, dieback results from root destruction. When the root system is suddenly diminished and becomes incapable of supporting the entire canopy, leaves wilt and twigs die back until a balance is restored between roots and tops. Roots may be destroyed by deep plowing, fluctuating water tables, drought, fertilizer burn or deficiencies, oil spillage, pathogenic organisms, and pests. To some extent, dieback also results from injuries to twigs and branches produced by hurricanes, prolonged hot, dry winds, salt-laden seaspray, and aerial pests.

The anthracnose fungus has also been held responsible for a russeting and tearstaining of the rind. Though this condition has been reproduced

by inoculating fruits with pure cultures of certain strains of *C. gloeosporioides* (16), practically all cases of russet and tearstain in Florida are now attributed to the citrus rust mite (366) and to melanose.

Another type of injury with which *C. gloeosporioides* has been associated is a spotting of the rind (276). Lesions are generally round, brown to black, sunken, hard, ⅛–¾ in. in diameter, and often dotted with the minute black fruiting structures of the fungus. In late stages, spots afford entry to other organisms that hasten involvement of the rind and rotting of the flesh. Anthracnose spots usually develop at sites of mechanical injuries or in fruits that are overripe. Grapefruits are particularly prone to the development of anthracnose spot. *C. gloeosporioides* has also been charged with producing a soft, pliable decay of the rind and the shedding of affected fruits.

Control of wither tip and anthracnose spot is afforded indirectly by measures that maintain tree vigor and that prevent fruit injury or senescence.

ARMILLARIA. Genus of the basidiomycetous fungi. *A. mellea* (Vahl) Kummer is one of the causal agents of **Mushroom Root Rot** (which see).

ARMILLARIA ROOT ROT. See Mushroom Root Rot.

ARSENIC TOXICITY. A foliar chlorosis resulting from excessive use of arsenic to reduce acidity in early harvested grapefruit.

Arsenic sprayed on grapefruit trees to lessen the acid content of fruit (83) may induce foliar chlorosis, especially if sprays are applied for a number of years in succession. Mildly affected leaves show a mottle in areas between and sometimes across the veins. Chlorotic spots extend to the leaf margins. The pattern resembles the one indicative of manganese deficiency except that arsenic chlorosis appears in mature leaves during summer and fall and usually commences on the south and southwest sides of trees. Leaves more severely affected develop an intense chlorosis or become yellow throughout the blade, followed by leaf drop. Fruit may be dwarfed, misshapen, hard, and spotted internally with gum—symptoms that resemble those of boron deficiency. In fact, excess applications of arsenic seem to induce boron deficiency, and applications of borax have been found to lessen the severity of arsenic toxicity symptoms (266). To promote recovery of trees, arsenic sprays should be omitted the year following occurrence of symptoms.

ASCHERSONIA. Genus of the deuteromycetous fungi. *A. aleyrodis* Webber, *A. goldiana* Sacc. & Ell., and *A. turbinata* Berk. are parasitic on whiteflies and scale insects. See **Entomogenous Fungi.**

ASPERGILLUS. Genus of the ascomycetous fungi. *A. niger* v. Tiegh., *A. alliaceus* Thom & Church, and *A. flavus* Lk. cause **Aspergillus Rot** (which see).

ASPERGILLUS ROT. A postharvest fruit rot of minor importance.

Among the miscellany of fungi encountered in citrus fruits that have been shipped or stored under conditions of high temperature are *Aspergillus niger* v. Tiegh., *A. alliaceus* Thom & Church, and *A. flavus* Lk. In fruit affected by a mixture of decay organisms including *Geotrichum candidum* Lk., aspergilli frequently predominate. The decay apparently spreads from affected to sound fruit by contact.

First symptoms following invasion are spots that are light-colored, watersoaked, and soft. In time, these areas become sunken, dark, and abundantly covered by black spore masses.

To prevent Aspergillus rot, fruit should be held in storage and shipped at recommended temperatures. Also applicable to control are the treatments given under **Blue Mold** and **Green Mold** (301).

BACTERIAL CANKER. See **Canker.**

BELONOLAIMUS. Genus of the phytopathogenic nematodes. *B. gracilis* Steiner (the sting nematode) has been proven to be pathogenic to citrus, but its economic effects remain to be assessed. See **Nematodes.**

BIG MEASLES. See **Measles.**

BIURET TOXICITY. A yellow blotching of foliage, resulting from the use of urea fertilizers containing biuret as an impurity.

Urea containing biuret produces a foliar chlorosis (Fig. 4) when applied to trees either through the soil or in sprays (246). Symptoms vary with biuret content: trace amounts affect only the tips; greater amounts involve the entire leaf blade except the midrib and may cause necrosis. The blotching resembles **Perchlorate Toxicity** (which see) but appears most prominently in young leaves, is cream-colored, and is not criss-crossed by green veins. Symptoms resemble also those of **Boron Toxicity** (which see) except for the absence of gummed areas on undersides of leaves.

No measures are known that will correct the chlorotic condition. Succeeding flushes will be normal if urea fertilizers free of biuret are substituted.

BLACK FUNGUS. *Myriangium duriaei* Mont. & Berk., one of the so-called "friendly fungi." See **Entomogenous Fungi.**

FIG. 4. *BIURET TOXICITY.* Chlorosis of leaf tip from traces of biuret in fertilizer. Yellowing may involve the entire leaf blade when the percentage of biuret is greater. (Photograph by F. P. Lawrence.)

BLACK MELANOSE. See Greasy Spot.

BLACK ROT. A decay of fruits on the tree, usually visible externally but at times detectable only after fruits have been cut open. SYN.—Alternaria rot, Alternaria spot, center rot, stem-end rot.

Black rot was formerly a disease of little importance, but with the advent of frozen concentrate, it has become a problem. The juicing of a few fruits with black rot may taint an entire pack. Although most fruits can be culled on the basis of external symptoms, those affected only internally cannot easily be eliminated.

Varieties frequently found affected in Florida are oranges (particularly Jaffa, Navel, and other early varieties), tangerines, and lemons.

Symptoms are most conspicuous a few weeks preceding normal color break. Affected fruits develop an orange color prematurely. Usually, these fruits show external evidence of infection in the form of a light-brown to blackish discoloration of the rind at or near the stylar end. A small percentage of fruits show no external evidence of infection and must be cut open for the diseased condition to be exposed. Internal symptoms consist of a brown to greenish-black discoloration and disorganization of the core, most pronounced at the stylar end (Fig. 5). Even a small amount of rot imparts a characteristic bitter flavor to the flesh. Affected fruits eventually shed.

Isolations from diseased tissues yield the fungus *Alternaria citri* Ellis & Pierce. It is not certain, however, whether such isolates can initiate the disease or whether they colonize tissues only after they have been weakened by other agencies. Fruits artificially inoculated with *A. citri* succumb to black rot only after a long incubation period (228). Prolonged periods of wet weather during the final stages of fruit maturation seem to increase the incidence of black rot.

Though copper sprays have been suggested for the control of black rot, no experimental evidence of their effectiveness exists. Some relief from the danger of tainting packs by the inadvertent inclusion of affected fruits may be obtained by delaying harvesting until all diseased fruits have fallen from the tree.

BLIGHT. A variously defined and poorly understood wilting and dieback of citrus trees. In common usage, the term blight is often applied to any decline that shows blight-like symptoms in the tops. SYN.—orange blight, limb blight, go-back, wilt, dry wilt, white wilt, leaf curl, roadside decline, Plant City disease.

Blight was first described in 1891, but despite continual investigation since then, its cause remains in doubt. The following discussion is restricted to blight as defined by Rhoads; it does not include

FIG. 5. *BLACK ROT.* Top: An apparently sound sweet-orange fruit. Bottom: Same fruit cut open reveals hidden infection.

two diseases of recent importance that are considered by some to be manifestations of blight under certain environmental conditions (see **Young Tree Decline, Sandhill Decline).**

Plants affected

Among scion varieties, specific mention has been made of the susceptibility of sweet orange, grapefruit, mandarin, tangelo, Temple, and kumquat, and among rootstock varieties, sour orange, sweet orange, and grapefruit, as well as seedling trees of these same varieties. Rough lemon, once thought to be unaffected (269), is now regarded as susceptible (57). An understanding of relative susceptibilities must await the testing of scion and stock varieties and their combinations under the same environmental conditions.

History and distribution

The term "blight" was first used in 1891 (333) to characterize a disease that had become prevalent in the counties of Putnam, Lake, Orange, Seminole, Volusia, Brevard, and Manatee. Following the freeze of 1894–95, the concern over blight lessened but increased again during the 1920s when trees planted after the freeze came of susceptible age.

Swingle and Webber (323) reported blight to be most prevalent in sandy hammocks, less prevalent in flatwoods, and least prevalent in sand hills. Rhoads (269) concluded that blight occurred most frequently in such soil series as Palm Beach, St. Lucie, Lakewood, Dade, and Gainesville (especially when underlaid by shallow rock formations) and occurred least frequently in soils having medium to poor drainage.

Blight today is seen much less frequently and appears to be restricted to older groves of the east coast and the Plant City area.

Importance

Estimates of annual losses due to blight have ranged from $150,000 during the last decade of the nineteenth century (323) to $1,400,000 during the early 1950s (57). All estimates, however, are suspect unless blight has been identified by the uncovering of intact roots during early stages of the disease. Once blight causes a decline of the top, roots starve and rot and become indistinguishable from roots destroyed by other agencies.

Symptoms

Most investigators agree that blight is characterized by three symptoms: wilting of the foliage despite adequacy of soil moisture; dieback that is not the result of obvious damage to trunk or roots; and postdecline emergence of watersprouts. No one of these symptoms, however, is specific; each can be caused by other agencies.

Distribution of blighted trees in a grove is random; trees adjacent to blighted specimens may remain normal for decades. Blight rarely attacks trees less than 12 years of age, though it may affect trees at any age thereafter. Once trees decline, they never recover. Occasionally trees decline rapidly and die, but generally they linger in a state of nonproductiveness.

First indications of blight are a dulling of the foliage followed by wilting. Symptoms may appear throughout the canopy or along a few branches, usually those near the ground. Wilting that commences on one side ultimately involves the whole canopy. At first, wilted leaves may recover overnight but later, wilt persists despite rainfall or irrigation. With time, leaves roll tightly, become lifeless, and drop, usually abscising between blades and petioles. Dieback sets in after leaves have shed. If attacks are sudden, involving the entire canopy from the start, leaves dry quickly, turn papery brown, and remain attached, a manifestation that has given rise to the term "white wilt" and that resembles the occasional sudden collapse brought on by tristeza.

Successive flushes in affected portions of the tree are retarded and dwarfed or may fail to develop. There is often a heavy and tardy bloom on trees with extensive dieback; fruit set, however, is poor.

With the start of the rainy season, there is a copious development of watersprouts from the upper trunk and lower branches of blighted trees. Sprout foliage is at first vigorous but later it, too, wilts and drops. Sprouting occurs more frequently in sweet orange trees than in grapefruit trees. Though long regarded as one of the symptoms of blight, sprouting seems a poor diagnostic aid since sprouts usually arise following canopy reduction from whatever cause (269).

In early stages of wilt, trees appear to have normal root systems (68, 269). This characteristic distinguishes blight from disorders with similar aboveground symptoms but with roots obviously damaged by water, excess fertilizer, nematodes, gophers, fuel oil spillage, or mushroom root rot.

No mention was made by Rhoads (269) of nutrient deficiency symptoms in affected trees, but Childs (57) reported symptoms in sprout leaves indicating deficiencies of zinc, manganese, and boron, which were not corrected by soil applications of these elements.

Cause

The cause of blight is not definitely known. The following possibilities, however, have been explored.

Pathogenic microorganisms.—Though Swingle and Webber (323), Rolfs (275), and Hume (158) considered blight to be contagious, Rhoads (269) was unable to transmit the disease by topworking some 165 healthy trees with budwood from affected trees,[1] by planting healthy trees in sites previously occupied by blighted trees, or by inoculating healthy trees with various fungi isolated from affected twigs, branches, trunks, and roots. Further attempts by Childs (57) and Cohen (68) also failed to demonstrate a transmissible agent. No correlation was found between blight and numbers or species of nematodes belonging to 18 genera (103).

Anatomical abnormalities.—Rhoads (269) found no discoloration or decay in the bark or woody cylinder of blighted trees. Cohen (68) could detect no significant differences in passage of air through stem and root pieces from blighted and nonblighted trees. Childs (64) showed the presence in blighted trees of a fungus which he identified as *Physoderma citri* and suggested that the mycelial mats of this fungus in twigs might cause a vascular obstruction that leads to the symptoms of blight.

Soil factors.—Various researchers have suggested that blight results from (a) a decay of roots on contact with hardpan and rock (269), (b) a low moisture-holding capacity of the soil (167), (c) an adverse pH reaction of the soil (57), and (d) poor cultural practices (269). Rhoads (269) stated that blight was a composite disorder with the most im-

1. Cohen (68) examined one of Rhoads' plantings 35 years after trees had been budded and still found no evidence of transmission.

portant variable being soil moisture (at times too much water but more often too little) and that the occurrence of blight was correlated with pockets of droughty hard-to-wet soil, pothole-bound roots, and layers of coquina rock close to the surface. The relative importance of each of these variables has not been tested; therefore, the nature of blight remains in doubt.

Physiological factors.—No significant differences have been found in the amino acid contents of blighted and nonblighted trees (68).

Control

It has been demonstrated that blighted trees cannot be restored to health by modified fertilization, insecticides, fungicides, pruning off of affected limbs, buckhorning, topworking to other varieties, supplemental irrigation, bark splitting, tree lifting, or the removal of affected trees to new locations (68, 269). At present, the only way of coping with blight is to remove stricken trees and to replace them with new trees.

BLIGHT-LIKE DISEASE. See Young-Tree Decline.

BLIND POCKET PSOROSIS. See Psorosis.

BLISTER ROT. See Blue Mold, Green Mold.

BLOSSOM BLIGHT. See Lime Anthracnose.

BLOSSOM-END DECLINE. See Endoxerosis.

BLOTCH. See Lime Blotch.

BLUE MOLD. One of the postharvest rots of citrus fruits; seldom seen in Florida. *SYN.*—blue contact mold, pinhole rot, blister rot.

Blue mold (caused by *Penicillium italicum* Wehmer) resembles **Green Mold** (which see) but differs in being characterized by the bluish color of the powdery growth, by the unwrinkled surface of the colony, by sporulation within the flesh as well as on the rind, by a narrow white mycelial border around the blue portion of the colony, and by the development of a definite ring of water-soaked tissue in advance of the white mycelial border. In contrast to green mold, it spreads to adjacent fruit by contact. Both molds can be controlled by the methods discussed under **Green Mold.**

BORON DEFICIENCY. See **Lumpy Rind.**

BORON TOXICITY. Damage to leaves and fruit from excessive applications of borax.

The range between too little boron and too much boron is quite narrow in citrus (266). Inadequate amounts produce fruits that are undersized, lumpy, hard, and spotted with gum on the rind and in the axils and albedo. Excessive amounts of boron are indicated by a tip and marginal leaf chlorosis and necrosis and by the formation of gum over affected areas on the undersurfaces of leaves. The chlorosis spreads throughout the leaf blade, involving areas between the main lateral veins. In severe cases, leaves are reduced in size and are shed. Leaves flushing immediately after toxic applications may be entirely chlorotic except for green bands bordering the midribs and main lateral veins. Succeeding flushes show a progressive diminution of yellowing.

Excessive amounts of boron applied to the soil can be leached out by heavy irrigation or rainfall, or can be neutralized by liming. The relation between boron and arsenic is discussed under **Arsenic Toxicity.**

BOTRYOSPHAERIA. See **Dothiorella.**

BREVIPALPUS. Genus of the false spider mites. *B. phoenicis* (Geijskes) causes **Brevipalpus Gall** (which see). *B. californicus* (Banks) is associated with **Leprosis** (which see). *B. obovatus* Donnadieu, the cause of leprosis in South America, has recently been found on citrus in Florida (201).

BREVIPALPUS GALL. A mite-induced galling at nodes of sweet and sour orange seedlings which may lead to death of plants in the nursery.

In Florida, most galls found on citrus are of genetic origin. One of the exceptions is a nodal galling that occurs at times on stems of sweet and sour

orange plants in the nursery. The cause is *Brevipalpus phoenicis* (Geijskes), a false spider mite. On plants defoliated by disease, cold, wind, or drought, mites congregate at nodes and feed on emerging pinpoint sprouts. Prolonged attacks lead to bud proliferation. In time, cushions of aborted sprouts along the stems become gall-like structures up to ¼ in. in diameter (Fig. 6). Prevention of re-foliation leads ultimately to death of plants. In Venezuela, up to 60 per cent of sour orange seedlings in nurseries have been destroyed in this manner (212).

Controlled experiments have verified the causal relation between this mite and Brevipalpus gall (200).

One spraying with wettable sulfur (5–10 lb./100 gal. water) or chlorobenzilate (¼ pint of 45.5 per cent liquid/100 gal.) will eradicate Brevipalpus mites and will permit plants to sprout normally and to recover.

BROMELIADS. See **Spanish Moss.**

FIG. 6. *BREVIPALPUS GALL.* Nodal cushions resulting from the repeated killing of successive sprouts on a sour-orange seedling. Inability to develop new foliage causes eventual death of attacked seedlings.

BROWN FUNGUS. *Agerita webberi* Fawc., one of the so-called "friendly fungi." See **Entomogenous Fungi.**

BROWN ROT. A tan, leathery decay of citrus fruits on the tree, in storage, and in transit.

The Phytophthora fungi that cause foot rot are also capable of causing brown rot (Fig. 7). Brown rot was first encountered in Florida on the east coast in 1951 (359); since then it has appeared in Hardee and Polk counties (179). Attacks are sporadic and usually of minor importance, but at times the disease may affect the entire crop of individual

FIG. 7. *BROWN ROT*. The darkened area is olive-brown in color and leathery in texture. The white mold is the causal fungus. (Photograph by J. O. Whiteside.)

trees. Outbreaks are more frequent on the east coast than in the interior.

Plants affected

All varieties of citrus are potentially susceptible, but so far only sweet orange, Temple, and grapefruit have been found affected in Florida. When attacks occur in late summer, early and midseason varieties may become infected while adjacent late varieties react as if they were immune. However, this apparent immunity is not a function of varietal resistance but of immaturity: young fruits are not susceptible. Valencia may become infected during occasional prolonged rains in winter.

Symptoms

The rot starts as a light-brown discoloration of the rind. Usually there is only one spot per fruit, but at times the rot may involve nearly the entire rind. The rotted area retains the same degree of firmness and elevation as that of the adjacent healthy rind, that is, until secondary organisms enter and reduce the rot to a soft consistency. A faint white tufty mold covers the rotted area during periods of high humidity. Fruits fall soon after attack. A large quantity of decaying fruits under the tree is presumptive evidence that brown rot is, or has been, active. The presence of brown rot in a grove can often be detected by the distinctive, pungent, aromatic odor emitted by affected fruits.

Occasionally the young leaves and twigs on affected trees are attacked. The resulting patches are at first water-soaked; later, on drying, they turn brown to black.

In the packinghouse, fruits affected by brown rot have much the same appearance as in the field. Incipient infections, not visible at time of grading, may develop into large lesions within a week of packing.

Cause

Although *Phytophthora nicotianae* B. de Haan var. *parasitica* (Dast.) Waterh. (the fungus commonly associated with foot rot) is capable of causing brown rot, the fungus most frequently encountered in Florida epidemics is *P. citrophthora* (Sm. & Sm.) Leonian (361). Elsewhere in the world, brown rot is also caused by *P. syringae* Kleb., *P. hibernalis* Carne, and *P. palmivora* Butler.

Serious outbreaks develop only after fruits have been continuously wet with dew or rain for approximately 18 hours. Durations of this length occur most frequently in late summer (179), particularly during long periods of rain accompanying hurricanes. At times, similar durations occur in winter.

Brown rot is initiated by the splashing of soil-borne zoospores onto low-hanging fruits. Zoospores germinate quickly and penetrate the intact rind. Given the maintenance of a moisture film, the length of time between inoculation and first visible symptoms of decay is 4 days or more, depending on temperature.

Once infection has occurred, the fungus sporulates and provides new inoculum that may be splashed to fruits successively higher up the tree. *P. citrophthora* produces sporangia more rapidly and in greater numbers than *P. nicotianae* var. *parasitica*, which may account for *P. citrophthora* being the dominant species encountered in Florida outbreaks of brown rot (361).

Control

In the field, brown rot can usually be curbed by spraying the lower 6 ft. of the tree with neutral copper (2 lb. metallic/500 gal. water). This protective spray should be applied in advance of expected rains at the time early season varieties approach maturity, about the middle of August. The expense is justified only in areas where brown rot has been a problem in the past. The heading up of skirts, the hedging of trees, and the chopping of cover crops will reduce the likelihood of infection and hasten the drying of moisture on the rind.

In the packinghouse, various measures can be taken to reduce the danger of further contamination (169), but under Florida conditions, the prevailing practice is to grade out the brown rot and to reconsign the sound fruit to the canning plant.

BROWN-ROT GUMMOSIS. See **Foot Rot.**

BROWN STEM. See **Stem-End Rind Breakdown.**

BUD-UNION DECLINE DISEASE. See **Tristeza.**

BURNT STEM. See **Stem-End Rind Breakdown.**

CACHEXIA. See **Xyloporosis.**

CALIFORNIA SCALY BARK. See **Psorosis.**

CANCROID SPOT. An apparently genetic disorder in Valencia sweet orange fruit, producing lesions resembling those of bacterial canker.

Though of slight economic importance, cancroid spot deserves mention because of its striking resemblance to bacterial canker. Lesions on fruit (Fig. 8A, B) develop the same corkiness and concentric pattern long associated with canker. The resemblance is greatest when cancroid spots approximate the size of canker lesions (⅛–¼ in. in diameter); with time, however, lesions increase to a diameter of ¾ in. when, because of their large size and eruptive form, they are no longer mistakable for canker.

Associated with fruit symptoms is an inconspicuous dotting of the foliage (Fig. 8C), designated glassy spot. Dots range in size from pinpoints to pinheads and are bordered by faint halos. When viewed with transmitted light, they are translucent; with reflected light, they appear as obscure grayish depressions on the undersides of leaves. From a few to hundreds of dots occur on a single leaf. Affected leaves abscise between blades and petioles and lead to twig dieback and thinning of the canopy. The trouble has been seen only in Valencia sweet orange trees.

Cancroid spot and glassy spot appear to result from a genetic aberration (185). In transmission trials, glassy spot recurred only in the foliage from affected buds. Seeds from cancroid-spotted fruit give rise to glassy-spotted seedlings. Affected trees should be avoided in the collection of budwood.

CANCROSIS. See **Canker.**

CANCROSIS B. A variant form of canker in Argentina. Its symptoms are those of true canker but its suscept range differs. See **Canker.**

CANDELOSPORA. Genus of the deuteromycetous fungi. *C. citri* Fawc. & Klotz has been isolated from a side rot of orange fruits (93). Rare.

CANKER. A destructive bacterial disease producing lesions on fruit, leaves, and twigs; once widespread in Florida but following eradication, not seen in the state since 1926. *SYN.*—bacterial canker, cancrosis.

Canker was originally introduced into the Gulf states about 1910 on nursery stock from Japan. Florida's humid growing conditions were particularly favorable for the spread and development of canker, and before eradication was completed at a cost of $6 million, 257,745 grove trees and 3,093,110 nursery plants were put to the torch.

Though canker no longer exists in the United

States, it is included here because the disease may, for two likely reasons, reappear in Florida: (a) due to congestion at international ports of entry, it is no longer practical to inspect 100 per cent of incoming baggage for fruits and budsticks, and (b) due to relaxation of Federal Citrus Quarantine 28, it is now permitted to import Japanese citrus into the states of Washington, Oregon, Montana, and Idaho. Only with the vigilance of all segments of Florida's citrus industry—growers, production managers, packinghouse personnel, and regulatory officials—could a reinvasion be detected and nipped in the bud, thereby preventing heavy losses to growers and costly eradication programs for the state.

Plants affected

Nearly all varieties of citrus are susceptible. Most affected is grapefruit followed by trifoliate orange, Key lime, sweet orange, lemon, and Satsuma mandarin. Resistant to infection are Dancy-type mandarins, calamondins, citrons, and kumquats.

Symptoms

Canker is seen commonly on young leaves (Fig. 9A), twigs (Fig. 9B), and fruits (Fig. 9C). Lesions appear first as small, slightly raised, watery, circular spots, usually darker green than the surrounding tissue. On leaves, spots occur predominantly on the lower surface. In time, lesions become grayish-

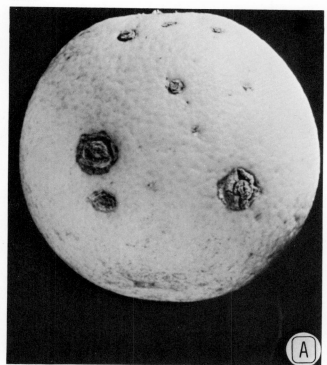

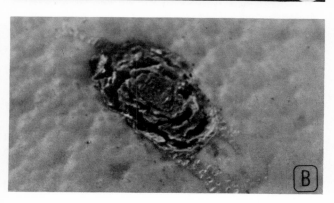

FIG. 8. *CANCROID SPOT.* A. Canker-like lesions on Valencia sweet orange. B. Cancroid spot enlarged. C. Glassy spot, a leaf symptom associated with cancroid spot.

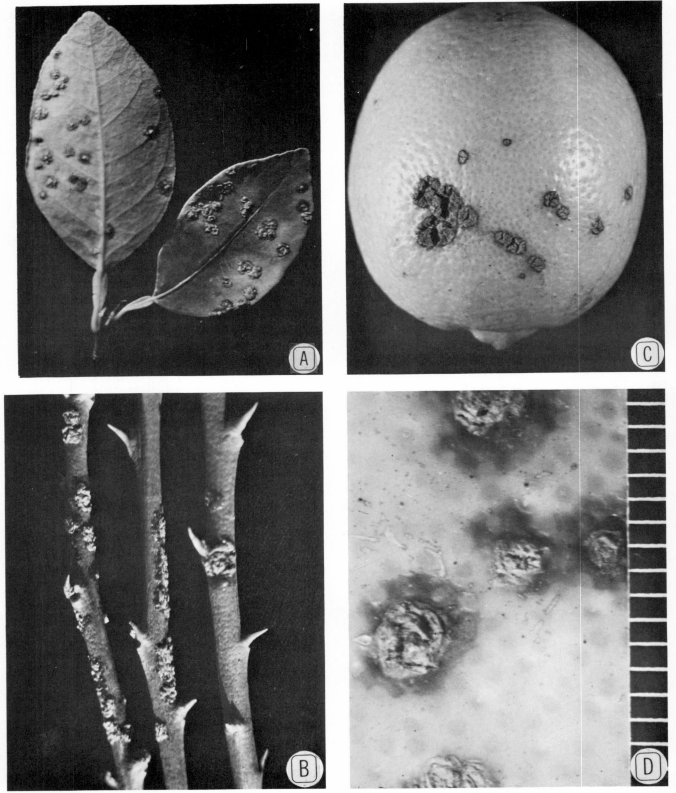

Fig. 9. *CANKER.* A. On leaves of kaghzi (Key-type) lime. B. On twigs of kaghzi lime. C. Cancrosis B on lemon. D. Enlargement of cancrosis B lesions on rind of sweet orange (with millimeter scale).

white and rupture to expose craters filled with concentric rings of tan, spongy, resinous, disorganized tissue (Fig. 9D). Surrounding each spot is a chlorotic halo that may persist until the lesion is quite old. Lesion size depends on variety affected; thus, on grapefruit foliage and fruit, spots may measure up to ½ in. in diameter, whereas on lemons and Key limes, spots generally do not exceed ⅛ in. Old lesions are brownish, but if overgrown by fungi, they assume other colors.

The development and appearance of canker on fruits and twigs are similar to those on leaves. At times lesions produced by lime anthracnose, scab, leprosis, and cancroid spot can be mistaken for those of canker (189).

Cause

Canker results from infection by *Xanthomonas citri* (Hasse) Dowson, a short, rod-shaped, motile, Gram-negative bacterium. Infection occurs when bacteria oozing out of canker lesions during rainy weather are splashed over young tissues and enter plant tissue by way of the stomates. In older tissues, ingress may also take place through injuries to the epidermis. Optimum conditions for infection occur between 68 and 95°F during periods when free moisture covers susceptible host tissue for 20 minutes or longer (256).

Another form of canker, knows as cancrosis B, occurs in Argentina (84). It differs principally in its host range, affecting most seriously lemons and to lesser degrees sweet and sour orange, grapefruit, lime, and citron.[1]

Control

Now that canker has been successfully eradicated from Florida, the first line of defense must be the exclusion of citrus from canker-infested areas of the world.

1. Recently, in Brazil, a third strain of the canker bacterium was found to differ serologically from the ones causing cancrosis A and B. The name *Xanthomonas citri* var. *aurantifolia* has been proposed for this third variant which affects mainly the small, acid, seedy lime (Namekata, T. 1971. Estudos comparativos entre Xanthomonas citri [Hasse] Dow., agente causal do "Cancro Citrico" e Xanthomonas citri [Hasse] Dow., n.f. sp. aurantifolia, agente causal da "Cancrose do Limoeiro Galego." Doctoral thesis, Escola Superior de Agricultura "Luiz de Queiroz," Piracicaba, Brazil).

Should canker again become established in Florida, it will have to be fought in the same manner as in the past: burning of infected trees by flame throwers, and disinfesting of workers' clothes and tools to prevent grove-to-grove spread.

CAPNODIUM. Genus of the ascomycetous fungi. *C. citri* Berk. & Desm. and *C. citricola* McAlp. are species commonly encountered in Florida producing **Sooty Mold** (which see).

CASSYTHA. Genus of the laurel family. One species, *C. filiformis* L., parasitizes citrus trees and forms a loose mantle of wire-like strands over the canopy. *SYN.*—dodder laurel, woe vine.

Many kinds of vines infest citrus trees. Those that are broad-leaved and manufacture their own food fall outside the scope of this publication. Cassytha and dodder are included here because they are parasitic, drawing their food and water requirements from the tree. Both are similar in appearance and are generally confused by all but botanists (149).

Cassytha filiformis L., commonly known as woe vine or dodder laurel, belongs to the laurel family whereas dodder belongs to the morning-glory family. Fruits of cassytha are small, 1-seeded drupes; those of dodder, capsules, mostly 4-seeded. When viewed under a hand lens, stems of cassytha are longitudinally striated, whereas those of dodder are smooth. Cassytha is a perennial, dodder an annual. Cassytha occurs in waste places of central and south Florida but is seen in citrus mostly on the east coast from Cocoa southward; dodder, on the other hand, occurs in fertile locations throughout the state.

Like dodder, cassytha starts from the ground as a seedling. Once in the canopy of a citrus tree, it severs connection with the ground and obtains food and water through suckers penetrating the bark of twigs. Subsequent ramification of the leafless, tendril-like, yellow-brown stems covers the canopy with what, at a distance, looks like a heavy matting of wire (Fig. 10). Longstanding infestations stunt the host.

If infestations are light, hand pulling provides satisfactory control; if heavy, hatracking of the tree is more economical.

FIG. 10. *CASSYTHA INFESTATION*. Cassytha *filiformis*, similar in appearance to dodder, parasitizing Key lime tree.

CENTER ROT. See **Black Rot.**

CEPHALEUROS. Genus of the Algae. One species, *C. virescens* Kunze, causes **Algal Disease** (which see). *SYN.*—*C. mycoidea* Karsten, *Mycoidea parasitica* Cunningham; possibly also *C. parasiticus* Karsten and *Phyllactidium tropicum* Mobius.

CEPHALOSPORIUM. Genus of the deuteromycetous fungi. *C. lecanii* Zimm. overgrows dead scale insects. See **Entomogenous Fungi.**

CERCOSPORA. Genus of the deuteromycetous fungi. *C. citri-grisea* Fisher has been described as causing **Greasy Spot** (which see). *C. gigantea* Fisher has been reported to cause tar spot of citrus leaves and fruit (106). *C. fumosa* Penz. has been isolated from necrotic leaf spots (93).

CHAETOTHYRIUM. Genus of the ascomycetous fungi. *C. hawaiiense* Mendoza has been found associated with a leafspot on grapefruit in Florida (5). Rare.

CHILLING INJURY. A storage and transit rind collapse of fruits, especially limes and grapefruit. *SYN.*—rind breakdown, rind pitting, scald.

Grapefruit and limes (both Key and Tahiti) often develop a rind collapse in storage or transit. Sunken portions of the rind are sharply delimited from adjacent normal portions (Fig. 11) and range in color from tan to dark brown. Symptoms resemble oleocellosis except that oil glands do not stand out prominently, areas of the affected rind are larger, and the breakdown occurs most frequently in the vicinity of the fruit equator. The disorder does not extend into the flesh but does render fruits more readily subject to decay by secondary fungi.

The rind breakdown is a form of chilling injury and is likely to develop, especially in waxed fruits, at temperatures below 50°F. Such temperatures are frequently employed to keep limes green and to check decay, but low temperatures accomplish these advantages at the risk of incurring rind breakdown.

For prevention, fruits should be held only for short periods at 50°, preferably under high humidities, and then shipped promptly (87, 255). See also **Oil-Gland Darkening.**

CHIMERIC BREAKDOWN. A sectorial chimera of the rind and a breakdown of internal tissues in fruits of Tahiti lime.

Bud mutations are common in Tahiti lime. One type, known as chimeric breakdown, is characterized by a sectoring of the rind and a disorganization of the flesh (198). The sectoring consists of approximately 10–15 chimeric bands extending from the stem end to the stylar end of the fruit. When sectored fruits are cut longitudinally, they reveal a gumming of the axils, segment walls, and inner

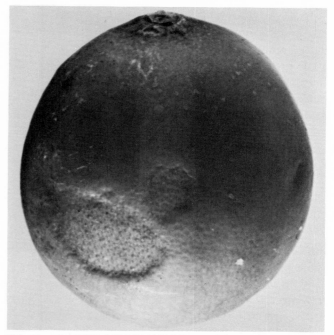

FIG. 11. *CHILLING INJURY.* Tahiti lime exhibiting low-temperature rind breakdown. Affected areas range in color from tan to dark brown.

albedo. The leaves on trees with affected fruits are narrower than those of the characteristic Tahiti lime. The disorder appears to result from a genetic aberration; as such, it could be controlled by avoiding affected trees during the collection of budwood for propagation.

CINNAMON FUNGUS. *Verticillium cinneamomeum* Petch, one of the so-called "friendly fungi." See **Entomogenous Fungi.**

CITRANGE STUNT. See **Tatter Leaf–Citrange Stunt.**

CITRUS-ROOT NEMATODE DISEASE. See **Slow Decline.**

CLADOSPORIUM. Genus of the deuteromycetous fungi. *C. citri* Massee was at one time thought to cause **Scab** (which see). *C. herbarum* Lk. var. *citricola* Farl. is associated with, and was at one time thought (90) to be the cause of, **Leprosis** (which see). *C. oxysporum* Berk. & Curt. has been found in necrotic tissue of **Cladosporium Leafspot** (which see).

CLADOSPORIUM LEAFSPOT. A leaf-spotting disease in the nursery.

Leaves on nursery trees of sweet orange, mandarin, grapefruit, lime, and lemon occasionally show tan to brown circular spots surrounded by dark brown, raised margins. With age, affected tissues degenerate to leave shothole effects. A fungus, *Cladosporium oxysporum* Berk. & Curt., has been isolated from the necrotic spots (108).

CLITOCYBE. Genus of the basidiomycetous fungi. *C. tabescens* (Scop.) Bres. is one of the causal agents of **Mushroom Root Rot** (which see).

CLITOCYBE ROOT ROT. See **Mushroom Root Rot.**

COLD DAMAGE. Low-temperature injuries to foliage, fruit, and bark.

Many factors influence the reactions of citrus to frosts (rapid losses of heat through radiation at night) and freezes (cold fronts displacing warm, moist maritime air). A discussion of these factors, as well as measures for their control (130, 213), falls outside the scope of this publication. The following remarks are limited to the symptomatology of frozen tissues.

Foliage responds to a slight freeze by developing a dark, greasy appearance and becoming flabby. Leaves so affected often recover. More severe damage leads to leaf curling, parching, and dropping. Defoliated limbs bear no fruit the following season.

The sequence of damage to the fruit as caused by increasing cold and duration are: (a) development of water-soaked areas on the membranes; (b) formation of white hesperidin crystals on the membranes and in the pulp; (c) appearance of an oleocellosis-like spotting of the rind, particularly in grapefruit, lemon, and lime; and (d) destruction of the water-retentive property of the rind so that excessive loss of moisture takes place, resulting in the drying out of the pulp and the separation of segments. Often there is no external evidence of inner destruction (311).

Bark damage in general varies with the diameter of twigs, branches, and trunks—the greater the diameter, the greater the tolerance to cold. The trunk is the last part of the tree to be affected. If trees are severely frozen, limbs die back, cold cankers develop on the trunks (particularly at or near the crotches), and the bark loosens or splits to allow entry of secondary decay organisms that initiate **Wood Rot** (which see). Usually bark damage does not become evident until weeks or months after freezes occur.

COLLAR ROT. See **Foot Rot.**

COLLETOTRICHUM. Genus of the deuteromycetous fungi. *C. gloeosporioides* Penz. (the imperfect stage of *Glomerella cingulata* [Stonem.] Spauld. & Schrenk) causes latent infections in healthy tissues and produces fruiting bodies in debilitated tissues. See **Alternaria Leafspot, Anthracnose, Symptomless Infections.**

CONCAVE GUM PSOROSIS. See **Psorosis.**

CONCENTRIC CANKER. See **Wood Rot.**

COPPER DEFICIENCY. See **Exanthema.**

COPPER TOXICITY. Injurious effects to trees from excess copper in the soil; also rind blemishes from improperly applied copper fungicides.

At one time, exanthema was widespread in Florida. With recognition that it was due to copper deficiency and that it could be corrected with supplemental copper, exanthema ceased to be important. However, in its wake a new problem arose: copper toxicity. Many older groves on acid, sandy soils are now in decline because of a chronic nutritional imbalance caused by copper that has accumulated after many years of fertilization and spraying (267). Adequate levels for the prevention of exanthema are approximately 50 lb. of total copper/acre 6 in. Some sandy, acid, grove soils now test as high as 600 lb. (266). When levels exceed 100 lb. (much higher in calcareous and organic soils), trees may begin to decline. Leaves may develop symptoms of iron deficiency, canopies become thin, tree growth is retarded, yields are reduced, and feeder roots become darkened, stubby, and sparse. Since these symptoms may be produced by other causes, their attribution to copper toxicity must be substantiated through determinations of total copper in the soil. Leaf analyses do not provide satisfactory indications of excess copper in the soil because this element does not always accumulate proportionately in leaves; however, values exceeding 15–20 ppm copper in dry tissue should be regarded with suspicion.

Excess copper can be rendered largely unavailable to citrus roots by liming the soil to a pH value above 6.5. This will usually lead within a year or so to great improvement in tree condition. Where high copper concentration is known to be a problem, copper should be omitted from fertilizers and trees should be sprayed, where possible, with non-copper-containing fungicides.

Copper applied as a spray may also cause a burn on the rind in addition to intensifying previous blemishes (see **Star Melanose**). Copper–lime-sulfur applied other than as a dormant spray may cause fruit spots that are tan to dark brown, sunken in the centers, and ringed by raised margins. Copper-oil burn appears as tan to gray, corky, roughly circular spots that are distributed over sides of the fruit where run-off forms drops that dry on the surface. For prevention of these injuries, copper mixed with lime-sulfur should not be applied other than as a dormant spray, and copper mixed with oil should not be applied later than 4 weeks following petal drop.

CORTICIUM. Genus of the basidiomycetous fungi. *C. stevensii* Burt (at times given as *C. koleroga* [Cooke] Hoehn.) is the cause of **Thread Blight** (which see).

CREASING. A disorder of fruit on the tree exhibiting furrowing and weakening of the rind. *SYN.*— crinkly skin, wrinkle-skin, puffing (erroneously).

Creasing is a rind disorder of sporadic occurrence, varying from year to year in degree, incidence, and distribution. It is characterized by diffuse irregular grooves on the surface of the fruit (Fig. 12). When affected fruits are squeezed, the peel separates along the furrows. Creasing results from a collapse of the rind over portions of the albedo where the cells have separated. It differs from **Puffing** (which see) in that the latter disorder is due to an irregular thickening of the rind. Creased fruits have thinner skins, higher percentages of juice, higher total soluble-solids-to-acid ratios, and higher specific gravities than noncreased

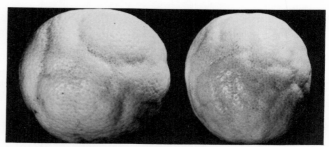

FIG. 12. *CREASING.* Settling of the rind over portions of the albedo where cells have separated. (Photograph by W. W. Jones.)

fruits from the same tree, indicating that creasing is a function of maturity. A high correlation generally exists between the incidence of creasing and the number of fruits per tree (165).

Creasing appears to be caused by a complex of factors including fruit exposure, temperature ranges, genetics, and nutrition (165). High soil applications of potassium reduce creasing (299) but increase other adverse fruit-quality effects. Potassium nitrate sprays lessen creasing at harvest even when applied during the presence of creased fruits on the tree (165).

CRINKLE-SCURF. A disorder of Valencia sweet orange trees characterized by a twisting and chlorosis of leaves and a horizontal banding of corky blisters on trunks and main limbs.

In this hyphenated term, "crinkle" refers to a twisting of leaf blades (Fig. 13A), conspicuous as soon as leaves expand and persisting through maturity. Each twisted leaf contains a central large chlorotic area within the blade, best seen by transmitted light (Fig. 14). The term "scurf" refers to closely spaced lines of corky pimples on the bark of trunks and main limbs (Fig. 13B). Crinkling and scurfing appear together in approximately 95 per cent of affected trees.

Crinkle-scurf causes significant reductions in trunk diameter, fruit size, yield, and juice content and a significant increase in soluble solids (265).

The incidence of crinkle-scurf in Florida varies from grove to grove. In most Valencia plantings, only an occasional instance of this disorder is seen; in other plantings, up to 54 per cent of trees are

FIG. 13. *CRINKLE-SCURF.* A. Leaf crinkling. B. Bark scurfing.

affected. Variations in incidence also occur within single trees; at one extreme the whole tree may show leaf symptoms, at the other, the foliage of only a single limb.

Formerly the disorder was thought to be a form of psorosis (95, 340), but recent studies (178) indicate that crinkle-scurf is a genetic abnormality.

Eradication or topworking of affected trees is recommended only in groves containing a high percentage of the disorder and in groves used as sources of budwood since crinkle-scurf can be propagated through budsticks from affected limbs.

CRINKLY LEAF. A virus disease causing puckering of leaves; of infrequent occurrence in Florida. Not to be confused with **Crinkle-Scurf** (which see).

Lemon and grapefruit trees occasionally show crinkling and puckering of leaves on one or more twigs or branches (Fig. 15). The symptoms are often conspicuous but damage to trees appears slight. In indicator seedlings, the initial crinkling reaction is sometimes followed by a pinpoint stippling.

The leaf puckering resembles that induced by some strains of the infectious variegation virus. This similarity has prompted a comparison of both viruses with respect to cross-protection reactions (343), to *in vitro* characteristics (136, 224), and to symptoms produced in mechanically inoculated herbaceous plants (224). The evidence suggests that crinkly leaf and infectious variegation are caused by strains of a single virus. The two strains manifest different symptoms when inoculated mechanically into bean seedlings of the varieties Bountiful, Geneva Market, Price, Satisfaction, and Tender Green. In these indicator plants, crinkly leaf produces no visible reaction whereas infectious variegation produces bright yellow vein chlorosis, interveinal mottling, and top necrosis (224).

In lemon seedlings inoculated with seed-transmitted crinkly leaf virus, young leaves sometimes develop the leaf flecking indicative of psorosis. This has suggested that the crinkly leaf–infectious variegation complex may belong to the psorosis virus group.

As in the case of **Infectious Variegation** (which see), crinkly leaf can be avoided by selecting budwood and seeds for propagation from nonaffected trees.

CRINKLY SKIN. See **Creasing.**

CRYSTALLIZATION. See **Granulation.**

CUSCUTA. Genus of the convolvulus family. *C. americana* L., *C. campestris* Yuncker, and *C. Bold-*

inghii Urb. are known to parasitize citrus in nurseries and groves in Florida. See **Dodder.**

DALDINIA. Genus of the basidiomycetous fungi. One species, *D. concentrica* (Bolt. ex Fr.) Ces. & DeNot., has been reported to cause **Wood Rot** (which see).

DAMPING OFF. A stem and root rot of seedlings in the seedbed.

Citrus, along with most cultivated plants, is susceptible during the young seedling stage to various fungi that cause damping off. The trouble occurs in patches of recently emerged seedlings. Spread in the area of affected seedlings is rapid during cool wet weather. Once seedlings pass through the succulent stage and become woody, they are resistant to further attacks of damping off.

Presence of the disease commands attention by the drooping of leaves. An examination of the wilted plants shows a girdling of stems at or above the soil surface (Fig. 16). Affected tissues are water-soaked or withered, constricted, and blackened. When removed from the soil, plants show a rotting of the roots.

Several fungi are capable of causing damping off. The predominant one is *Pellicularia filamentosa* (Pat.) Rogers, more commonly known by the name of its imperfect stage, *Rhizoctonia solani* Kühn. Other pathogenic species reported from Florida are *Sclerotium rolfsii* Sacc., *Phytophthora* spp., and *Pythium* spp.

Conducive to the development of this disease are poor drainage, excessive watering, heavy shading, and overcrowding of seedlings (358). Damping-off organisms are common inhabitants of the soil and may also be seed-borne.

The easiest way to avoid damping off is to select well-drained soil for the seedbed and to fumigate the site before planting. Hot water treatment of the seed (29) will reduce the probability of seed-borne contamination.

Soil sterilization may be accomplished by fumigating with methyl bromide, Vapam, or chloro-

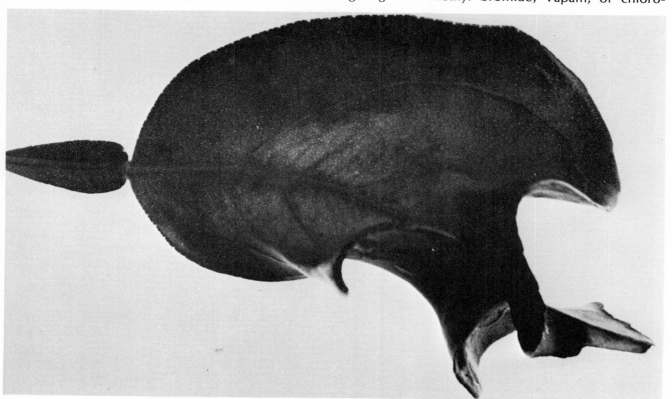

FIG. 14. *CRINKLE-SCURF.* Faint chlorosis within leaf blade best seen with transmitted light. (Photograph by Edgar Cary.)

picrin. Rates of application and duration of waiting period before planting are as directed on manufacturers' labels for preplant seedbed sterilization.

Should damping off appear in the seedbed, its further spread may be checked by grubbing out affected seedlings and by spraying the soil and plants with a fixed-copper compound at the strength of ¾ lb. metallic content/100 gal. water.

Judicious watering of the seedbed will do much toward preventing damping off. Irrigation should not be continued beyond wetting of the root zone, and the soil surface should be allowed to dry out until seedlings show the need for further watering. Such precautions are particularly important during cool cloudy weather.

DECORTICOSIS. See Shell Bark.

DENDROPOGON. Genus of bromeliaceous air plants. One species, *D. (=Tillandsia) usneoides* L., is known commonly as **Spanish Moss** (which see).

DEVIL'S GUTS. See **Dodder.**

DIAPORTHE. Genus of the ascomycetous fungi. *D. citri* (Fawc.) Wolf (the perfect stage of *Phomopsis citri* Fawc.) is the cause of **Melanose** (which see).[1]

DIEBACK. A term of two meanings: (1) specifically,

1. According to F. E. Fisher (Mycologia **44**:422. 1972), *D. medusaea* Nitschke has nomenclatorial precedence over *D. citri* and is to be preferred according to the *International Rules of Botanical Nomenclature*. Similarly, *Phomopsis cytosporella* Penz. & Sacc. takes precedence over *P. citri*.

FIG. 15. *CRINKLY LEAF*. A puckering of grapefruit leaves induced by the citrus crinkly leaf virus.

a synonym for **Exanthema** (which see); (2) generally, any death of twigs and branch terminals without respect to cause.

DIPLODIA. Genus of the deuteromycetous fungi. *D. natalensis* P. Evans (the imperfect stage of *Physalospora rhodina* [Berk. & Curt.] Cke.) causes a decay of fruits in the packinghouse (see **Diplodia Stem-End Rot**). It is reported to be the cause of Diplodia twig dieback (see **Robinson Dieback**). The

FIG. 16. *DAMPING OFF.* Stem girdling and root rotting of plants in the seedbed.

fungus can be found in and on healthy twigs and roots without causing symptoms (98, 277). It readily colonizes weakened tissue in which it produces fruiting bodies.

DIPLODIA GUMMOSIS. See Rio Grande Gummosis.

DIPLODIA ROOT ROT. See Water Damage.

DIPLODIA STEM-END ROT. One of several postharvest fruit rots originating at the stem end. *SYN.—* stem-end rot (a term also applied to similar rots caused by Phomopsis and Alternaria fungi).

Diplodia stem-end rot is one of the two most important fungus diseases of fruit in the packinghouse and in transit. The other is Phomopsis stem-end rot. Both rots are indistinguishable during early stages of decay and can be identified only on isolation of the causal fungi. Both diseases are usually combined under the name stem-end rot, and control procedures in the packinghouse are identical.

Symptoms

As in the case of Phomopsis stem-end rot, the earliest symptom is a softening of the rind around the button. The affected area increases rapidly and becomes brown. The discoloration is usually darker than that produced by Phomopsis. Fingers of decayed tissue extend down the rind in advance of the main body of the rot. A rotting of the rind may develop at the stylar end as a result of the more rapid progression of the internal decay (Fig. 17). This tendency to show decay at both ends is a characteristic of Diplodia stem-end rot.

Within the fruit, the rot progresses rapidly through the core and along the inner wall of the peel, eventually invading the flesh and imparting a bitter taste.

Peak decay occurs 2–3 weeks after harvest. Diplodia is the major cause of stem-end rot in fruits degreened with ethylene, whereas Phomopsis is the more usual cause of stem-end rot in fruits harvested later in the year and not degreened.

Cause

The causal fungus is *Diplodia natalensis* P. Evans, the ascigerous stage being *Physalospora rhodina* (Berk. & Curt.) Cke. Spores originate in dead wood in the tops of trees and are washed over fruits during rains. Spores that lodge between the calyx lobes and the rind are protected from desiccation and remain ready to establish decay. The major entryway is through breaks that result from abscission of the button (35). Fruits are not usually attacked while on the tree unless injured or overripe (234).

Control

The control of Diplodia stem-end rot in the packinghouse is the same as that given for **Phomopsis Stem-End Rot** (which see).

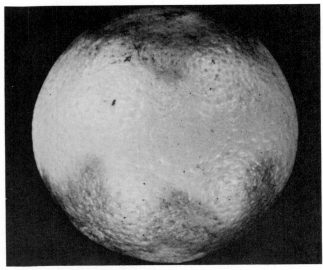

FIG. 17. *DIPLODIA STEM-END ROT.* The appearance of decay at both ends of the fruit during late stages of infection is a characteristic of this disease and differentiates it from the one-sided decay produced by Phomopsis stem-end rot. (Photograph by G. E. Brown.)

DIPLODIA TWIG DIEBACK. See Robinson Dieback.

DODDER. Genus of the convolvulus family, several species of which parasitize citrus trees. *SYN.—*devil's guts, gold thread, hellbind, love vine, strangle weed, vermicella.

Dodders are leafless, chlorophyll-less, flowering plants that derive food by parasitizing other plants. An infestation usually originates from seed. On emerging from the soil, the thread-like seedling twines around weeds which at first serve as hosts and then as ladders for the dodder to reach the skirts of citrus trees. Once the seedling finds a host, it loses connection with the ground. On citrus, the dodder tip encircles the stem, penetrates

the bark with approximately 8 root-like suckers per turn, and then repeats the process on encountering new twigs. Infested trees appear to be covered with yellow-orange string.

In trees of lemon and lime, penetrating suckers stimulate the development of spirally arranged galls, in size up to twice the diameter of the twig. Growth of twigs above the galls is stunted (174).

Several species of dodder are found in Florida on grove trees and nursery plants. *Cuscuta americana* L., a thick-stemmed dodder, is the most common (Fig. 18B). Occasional infestations are encountered of two thin-stemmed species: *C. campestris* Yuncker and *C. Boldinghii* Urb. (175).

Fortunately, dodder occurs infrequently on Florida citrus, but when present, its proper removal may take 48 man-hours per tree. The entire length of each branch must be examined and rubbed free of the tightly encircling dodder bracelets or else remaining fragments will sprout and reinfest the tree (Fig. 18A). In orange, grapefruit, and mandarin trees, dodder is killed annually by low temperatures. In lemon and lime trees, however, the suckers overwinter within the bark and sprout on return of warm weather; control, therefore, requires the hatracking of infested trees. In addition to stripping and hatracking, it is also necessary periodically to cultivate the ground beneath the canopy to destroy dodder seedlings that arise from previous infestations.

Some of the newer herbicides promise easy control but until these have been tested on citrus, the recommendations remain that dodder be removed by hand and that seedlings be kept down by cultivation.

A similar-appearing but unrelated parasite on citrus is discussed and differentiated under **Cassytha.**

DODDER LAUREL. See Cassytha.

DOTHIORELLA. Genus of the deuteromycetous fungi. *D. ribis* (Fckl.) Sacc. (of which the perfect stage is *Botryosphaeria ribis* Gros. & Duggar) has been reported (93) to cause a postharvest fruit rot nearly identical to **Phomopsis Stem-End Rot** and **Diplodia Stem-End Rot** (which see).

DRY ROT OF FRUIT. See Nematospora.

DRY TIP. See **Endoxerosis.**

DRY WILT. See **Blight.**

ELSINOË. Genus of the ascomycetous fungi. *E. faw-cetti* Bitanc. & Jenkins (the perfect stage of *Sphace-loma fawcetti* Jenkins) is the cause in Florida of **Scab** (which see).

ENDOXEROSIS. An internal breakdown in fruits of

yellowing near the stylar end, but yellow fruits give no external clue to the presence of disorganized flesh. As in the case of **Stylar-End Break-down,** the cause is presumed to be water stress, and the control, frequent irrigation. Some reduction in the amount of endoxerosis is obtained by early harvesting.

ENTOMOGENOUS FUNGI. Various fungi that feed on living or dead insects and mites, and produce mold-like colonies on citrus.[1]

FIG. 18. *DODDER*. A. Dodder *(Cuscuta americana* L.) regenerating from haustoria in the bark of a lime twig 20 days after stripping of vines from the tree. B. Flowering and seed production of dodder on a citrus shoot.

limes and lemons. *SYN.*—internal decline, blossom-end decline, dry tip, yellow tip.

Endoxerosis is an internal breakdown of lime and lemon fruits (17). The disorder is characterized by the presence of gum in the core, along the segment walls, and in the inner albedo. Symptoms are similar to those described under **Chimeric Breakdown** except for the absence of rind sectoring. Green fruits occasionally show an external

In warm, moist climates, various fungi abound that feed on living or dead insects and mites. A discussion of the role these so-called friendly fungi play in pest control (233) falls outside the scope of this publication. The organisms involved

1. Strictly speaking, entomogenous fungi are those that feed on live, not dead, arthropods. Some of the fungi mentioned here are now known to be saprophagous. They are included because at one time they were regarded as entomogenous.

are mentioned here merely to emphasize that, despite their often conspicuous presence (Fig. 19), they are harmless to citrus trees.

Fungi that are found on citrus and attack whiteflies are the Red Aschersonia (*Aschersonia aleyrodis* Webber), the Yellow Aschersonia (*A. goldiana* Sacc. & Ell.), and the Brown Fungus (*Aegerita webberi* Fawc.). Mealybug populations are reduced by *Entomophthora fumosa* Speare. Dead scale insects are overgrown by the Red Fungus (*Sphaerostilbe aurantiicola* [B. & B.] Petch), the Gray Fungus (*Podonectria coccicola* [Ell. & Ev.] Petch), the Black Fungus (*Myriangium duriaei* Mont. & Berk.), the Cinnamon Fungus (*Verticillium cinneamomeum* Petch), the Pink Fungus (*Nectria diploa* Berk. & Curt.), *N. coccophila* (Tul.) Wr., *Cephalosporium lecanii* Zimm., and *Myriangium floridanum* Hoehn. Other fungi parasitic on insects and mites are *Myiophagus ucrainicus* (Wise) Sparrow, *Hirsutella besseyi* Fisher, *H. thompsonii* Fisher, *Entomophthora fresenii* Nowa., and *E. floridanus* Weiser & Muma (233).

ENTOMOPHTHORA. Genus of the phycomycetous fungi. *E. fumosa* Speare is parasitic on mealybug, *E. fresenii* Nowa. on aphids, and *E. floridanus* Weiser & Muma on Texas citrus mite. See **Entomogenous Fungi.**

ETHYLENE BURN. See **Gas Burn.**

EUTYPELLA. Genus of the ascomycetous fungi. *E. citricola* Speg. has been reported to occur on dead twigs (372). It may be interspersed and possibly confused with *Diaporthe citri* (Fawc.) Wolf, the cause of **Phomopsis Stem-End Rot** and **Melanose** (which see).

EXANTHEMA. Fruit, twig, and leaf symptoms resulting from the lack of available copper in the soil. *SYN.*—dieback, ammoniation, red rust, copper deficiency.

Before the discovery of its cause, exanthema was a widespread problem in Florida citrus groves (111). Now that the corrective is known (144) and is routinely applied, exanthema is seen only occasionally.

Fruit symptoms consist of irregular dark brown gummy patches on the rind, gum pockets within the peel, and gum around the seed. Affected fruits are low in acid and juice content.

Twig symptoms include internodal swellings or gum pockets on young shoots, coatings of gum on terminal growth, S-shaped twisting of shoots, formation of multiple buds, and sprouting of several shoots from single buds. In severe cases, twigs die back.

Leaf symptoms include gum staining of the blades and development of abnormally large leaves. Multiple sprouting of buds gives trees a bushy appearance.

Copper deficiency is not seen in groves that have received recommended rates of copper in the fertilizer or that have been sprayed with copper-containing fungicides. Exanthema does not occur in soils that contain 50 lb. total copper/acre 6 in., and

FIG. 19. *ENTOMOGENOUS FUNGI.* The Red Aschersonia fungus on leaves of lemon.

such soils should not receive additional copper in the fertilizer. More damage is encountered today from an excess of copper in the soil than from its previous deficiency. Spraying with copper fungicides for many years has built up levels to the extent that some older groves are now in decline because of **Copper Toxicity** (which see).

Should symptoms of exanthema appear, as they sometimes do in young trees on highly organic soils, they can be corrected quickly by an application of the same copper-containing spray that is recommended for the control of melanose. In severe cases, two or three such sprays during the year may be required.

EXOCORTIS. A virus disease causing bark scaling and stunting of trees on certain rootstock varieties. *SYN.*—scaly butt.

The presence of exocortis in Florida was first reported in 1954 (203), and an indexing of random trees in the state in 1956 indicated the likelihood that the majority of trees in Florida were infected (237). At that time, exocortis did not affect production because few trees were on susceptible rootstocks. Today, however, the threat is greater because stocks like trifoliate orange and various of the citranges have become popular. In those areas of the world where large acreages are on susceptible rootstocks, exocortis has been more destructive than any other virus disease of citrus except possibly tristeza.

Exocortis causes tree stunting and bark scaling on trunks of susceptible stocks with a resultant decrease in production. Affected trees seldom die, but stunting may be so severe that sooner or later stricken trees have to be replaced.

Plants affected

Commercial varieties that are susceptible to exocortis are trifoliate orange, some citranges (e.g., Troyer and Carrizo), Rangpur lime, and sweet lime. Also susceptible are such varieties of minor importance as sweet lemon, Cuban shaddock, Sziwuikom mandarin, citron, and some of the other mandarin-lime hybrids (39, 226, 254, 287, 351). The lime bark disease of Tahiti lime has been attributed to the exocortis virus (288).

Even in varieties that have long been considered

tolerant, the virus may also be detrimental. Stunting can occur when infected lemon trees are grown on sweet orange, grapefruit, or sour orange (44, 45).

Exocortis virus also produces symptoms in petunias and velvet plants (*Gynura aurantiaca* and *G. sarmentosa*) when experimentally inoculated (350, 352).

Varieties requiring frequent thinning, e.g., lemons, have high rates of infection, probably because of the tree-to-tree spread of exocortis virus on pruning tools (2).

Symptoms

In groves, exocortis is recognized by the scaling that it produces on the trunks and crown roots of susceptible rootstocks (Fig. 20). Scaling does not usually appear before trees are 3 years old, and it may not commence until they are 8 years old or more. First symptoms consist of shallow, longitudinal splits in the bark below the bud union. In time, splitting occurs over the entire circumference of the rootstock portion of the trunk, and strips of bark up to 6 in. long and ⅛–¼ in. thick begin to scale off. Scaling extends into the crown roots belowground, but it is not conspicuous here because the flaking bark is rapidly decomposed by soil organisms.

Exocortis-induced scaling may at times be confused with scaling due to shell bark in lemon trees, to a genetic abnormality in trifoliate orange (see **Laminate Shelling**), and to podagra in the rough lemon rootstock portion of kumquat trees (209).

Another symptom of exocortis is tree stunting which may occur in the presence or absence of scaling. Weak strains of the virus that cause neither scaling nor stunting have been reported (46).

Symptoms appearing in indicator seedings are described under *Cause*.

A phloroglucinol color test has been devised to identify infection in the phloem ray cells of trifoliate orange (65), but this test is now largely supplanted by the citron test (see *Cause*).

Infected plants may also be detected by chromatographic methods that register marked increases of scopoletin and umbelliferone in the stock bark of trifoliate orange (99) and in the midribs and bark of citron (102). A threefold increase of free phenolic compounds in diseased trees on trifoliate orange as compared with comparable healthy trees has also been proposed as a means of identifying infected trees (100). Indexing on hypersensitive clones of citron Etrog, however, is so specific and rapid that biochemical tests, at least as presently devised, appear to be impractical for routine detection (294).

Cause

The cause of exocortis is an atypical plant virus which appears to exist, at least in part, as free ribonucleic acid (295). Strains have been found that range from virulent to mild (39, 46). Host reactions to these strains vary also with environmental factors; thus, high rates of nitrogen and phosphate fertilizers accelerate scaling (354).

The virus is surprisingly resistant to aging, to heat, and to various sterilants (126, 273). In addition to budding with infected wood, transmission may occur in the nursery from virus-contaminated hands or when budding knives and pruning tools have cut through infected wood and are then used on noninfected plants (124, 273). Tools may remain contaminated for at least 8 days (2), and they are

FIG. 20. *EXOCORTIS.* A. Scaling of the bark in the rootstock portion of a tree on trifoliate orange. B. The scaling may occur above the bud union if the scion variety is susceptible. Shown here is a citrange top on rough lemon root.

not reliably sterilized by heat or most disinfectants. Citrus varieties react differently to inoculation with contaminated tools; prone to infection are trifoliate orange, Carrizo and Morton citranges, Rangpur lime, Eureka lemon, and citron Etrog; less prone are sweet and sour oranges, rough lemon, and West Indian lime; and least prone are Orlando tangelo, Rusk citrange, and Duncan grapefruit (127).

Transmission has also been accomplished by dodder (350), but seed transmission has not been found to take place (19, 39, 117, 124).

Indexing for the presence of exocortis virus in suspect trees may be carried out on various indicator seedlings. Bark scaling symptoms appear in trifoliate orange, Rangpur lime, and sweet lime within 1–8 years, but yellow blotching of shoots may develop within 6 months (293). These hosts have been supplanted, however, by certain hypersensitive clones of citron (e.g., USDCS 60-13, OES-7, Arizona 861, and Beerhalter Corsican) that produce results within 3–26 weeks (39, 123).

Control

Trees affected by exocortis can be made to recover by inarching or by inducing the scion to root, but these expedients for by-passing the susceptible stock are tedious and expensive and cannot be recommended except for the preservation of specimen trees.

Exocortis can be avoided either by the use of tolerant stocks (see *Plants affected*) or, if susceptible stocks are to be used, by grafting them with virus-free budwood. Exocortis-free budwood source trees must be protected from contamination.

Because exocortis virus may be spread on contaminated budding knives and pruning shears, such tools must be sterilized between budsticks and during movement from tree to tree. Two disinfectants that have been found effective against the virus are (1) an aqueous solution of 2 per cent sodium hydroxide and 2 per cent formalin and (2) a 5 per cent solution of Clorox (273). The spread of exocortis virus during hedging and topping operations has not yet been proved but undoubtedly can occur.

Under experimental conditions, exocortis-free propagative material has been obtained by subjecting infected budwood source trees for 230 days to a temperature of 100°F and then selecting shoot tips for propagation (315).

Because declining trees cannot be restored to health, they should be replaced when no longer productive.

FALSE LEPROSIS. A disorder of Valencia sweet orange trees causing rind and leaf symptoms similar to those of leprosis.

Fruits and leaves are occasionally spotted in a manner suggestive of leprosis (Fig. 21). False leprosis is indicated if Valencia sweet orange is involved (a variety not affected by leprosis) and bark symptoms are absent. The cause of false leprosis has not been determined, but it is known that Brevipalpus mites are not responsible (201). Trees affected one year return to normal the next.

FELT. Membranous coverings on branches, leaves, and fruit resulting from the growth of certain non-pathogenic basidiomycetous fungi. *SYN.*—felt fungus, mompa, plaster disease.

Under humid conditions, citrus trees teem with epiphytes. Bark, leaves, and fruits support multitudes of slime molds, bacteria, fungi, algae, lichens, mosses, ferns, and spermatophytes. Most of these organisms are inconspicuously small, and nearly all except a few pathogenic fungi are harmless. Only the larger ones compel attention; these, too, are usually harmless.

Among the more readily visible forms is felt, a coating produced by certain basidiomycetous fungi. In the Gulf states, two species are encountered on citrus, *Septobasidium pseudopedicellatum* Burt and *S. lepidosaphes* Couch. Elsewhere in the world at least 17 other species, belonging to the genera *Septobasidium, Helicobasidium, Multipatina,* and *Anthina,* are known to produce similar structures on citrus.

S. pseudopedicellatum, the species commonly seen in Florida, forms a thick, puffy, gray or tan sheath around twigs and branches, sometimes extending out onto the base of leaves, pedicels, and fruits. The growth may completely cover several inches of stem (Fig. 22). The surface is either velvety or membranaceous, the interior soft and spongy. The fungus does not penetrate host tissue but colonizes scale insects.

Because of its harmlessness, control measures are not warranted. If felt appears offensive, it can be pruned off or rubbed away, and further infestations can be prevented by the same copper sprays that are used to control melanose, scab, and greasy spot.

FERMENT GUM DISEASE. See **Rio Grande Gummosis.**

FIRING. A sudden wilting and parching of foliage on certain limbs; attributed to various agencies.

Sudden wilting and killing of foliage in restricted portions of trees may be the result of blight, root rot, certain virus diseases, and trunk, branch, or root girdling or breakage (159). Firing is a particular type that appears after periods of drought or sustained dry winds. It is thought to be aggravated by injuries to the foliage from high populations of citrus rust mite, citrus red mite, and Texas citrus mite. It is seen most frequently in sand-soaked areas of a grove and is associated with **Mesophyll Collapse** (which see).

Firing may be differentiated from some of the other types of wilting in that parched leaves remain on the tree, affected twigs usually stay green, and the sides of trees involved are those exposed to prevailing winds.

Preventive measures that have been suggested are the maintenance of proper tree nutrition and soil moisture and the control of mites.

FLATWOODS DECLINE. See **Young-Tree Decline.**

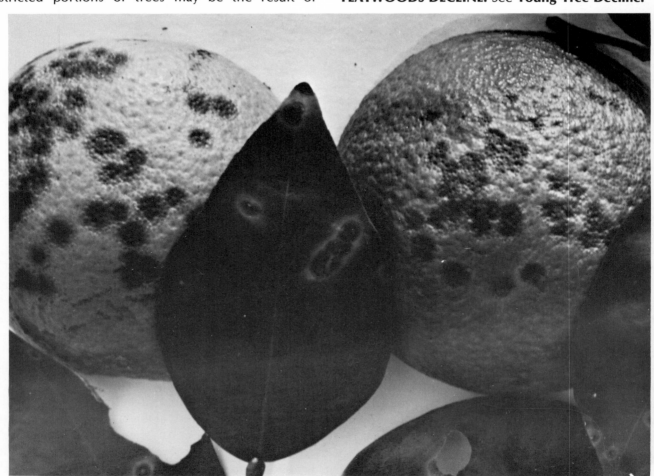

FIG. 21. *FALSE LEPROSIS.* Spotting on fruit and leaves of Valencia sweet orange.

FLORIDA GUMMOSIS. See **Rio Grande Gummosis.**

FLORIDA MOSS. See **Spanish Moss.**

FLORIDA SCALY BARK. See **Leprosis.**

FLUORIDE TOXICITY. Foliar chlorosis, tree stunting, and crop reduction due to atmospheric pollution with gaseous fluorine compounds.

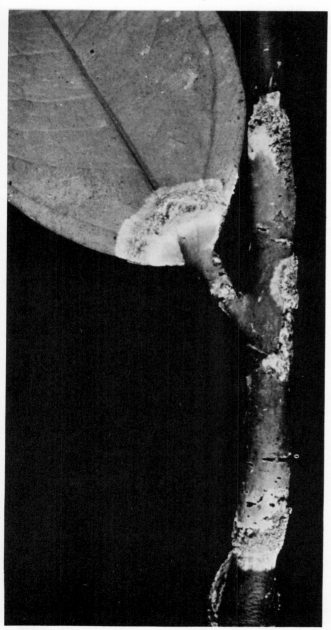

FIG. 22. *FELT.* Growth of *Septobasidium pseudopedicellatum* Burt on twig and leaf of grapefruit.

Fluoride fumes from phosphate fertilizer manufacturing plants may cause chlorosis, necrosis, and size reduction of leaves, stunting of trees, and lowering of fruit yield. Degree of damage depends on the concentration of fluoride gases and on the duration of exposure. Young leaves may develop chlorosis when they contain as little as 20 ppm of fluorine. Fruit production is decreased most severely when high levels of fluoride gas pollution coincide with the spring bloom (214, 215).

The foliar chlorosis is most prominent at the leaf tips and margins, and decreases toward the leaf bases (Fig. 23). On very young leaves, the chlorotic pattern often resembles that produced by manganese deficiency. As affected leaves become older, the manganese-deficiency-like pattern changes to the "Christmas tree" pattern shown in Figure 23. At times, fluoride toxicity chlorosis resembles that of boron toxicity except for the absence of gum on the undersurfaces of affected areas.

FLYSPECK. A minor rind blemish consisting of an aggregation of pinpoint-sized black specks.

In humid situations, many fungi proliferate on the rind of citrus fruits and produce grade-lowering blemishes. One fungus of minor importance is *Leptothyrium pomi* (Mont. & Fr.) Sacc., the cause of flyspeck. Symptoms consist of minute black specks most readily visible when congregated in patches. Each speck gives rise to pycnidia from which are extruded conidia that wash over other fruits during rains. The mycelium is incapable of penetrating the rind. Flyspeck is often found in areas of **Sooty Blotch** (which see). In contrast to sooty blotch, flyspeck is not easily removed by washing. The flyspeck fungus is susceptible to copper sprays, but control measures are seldom warranted.

FOMES. Genus of the basidiomycetous fungi. One species, *F. applanatus* (Pers.) Wallr., has been reported to cause **Wood Rot** (which see).

FOOT ROT. A rot of the trunk and crown caused by the same water mold that produces a rot of main and fibrous roots, a blight of plants in the nursery, and a decay of fruit. *SYN.*—Phytophthora gummosis, brown-rot gummosis, collar rot, mal di gomma.

At one time, when citrus fruits were produced on seedling trees, foot rot was the factor limiting yield. Following the discovery that the trouble

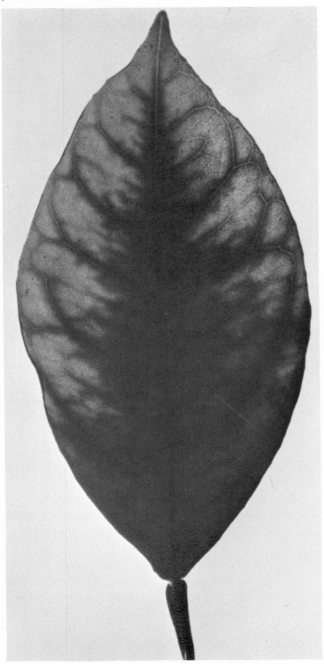

FIG. 23. *FLUORIDE TOXICITY.* Tip and marginal chlorosis against which green veins form a "Christmas-tree" pattern. (Photograph by C. D. Leonard.)

could be overcome by budding on tolerant rootstocks, foot rot became much less of a problem.

The rotting of bark near the ground line is only one of the destructive capabilities of the causal fungus. Other diseases caused by the fungus are a decay of fruits (see **Brown Rot**), a rot of fibrous roots, a frog-eye rot of main roots, and a blight of plants in the nursery. In Florida, these diseases are of minor importance compared with foot rot.

At one time, foot rot was more serious in bearing trees than in young trees (272), but in recent years it has become a problem in new groves during the first few years after planting (70).

Plants affected

Citrus species vary considerably in their susceptibility to foot rot, and this fact forms the basis for the major disease control recommendation. The results of many trials have shown the order of increasing resistance to be: smooth lemon, sweet orange, rough lemon, Cleopatra mandarin, *Citrus macrophylla,* Carrizo and Troyer citrange, and King, *C. taiwanica,* and various of the trifoliate oranges (170). This ranking, as confirmed by various investigators (50, 94, 115), is based mainly on resistance to feeder root infection rather than to foot rot, and resistance to *Phytophthora citrophthora,* a species not found in Florida foot rot lesions (70, 363). A study in Brazil, however, shows that the ranking of varieties is much the same whether the causal fungus is *P. citrophthora* or *P. parasitica* (94).

Though few trials to determine varietal susceptibility to foot rot have been conducted in Florida, field experience has led to the following ranking of rootstock varieties: highly susceptible—smooth lemon, Page, sweet orange, and grapefruit; economically tolerant—rough lemon, Cleopatra mandarin, Rangpur lime, sour orange, and various of the trifoliate oranges and citranges (184, 272, 363). Within any one variety, there may also be a range of reactions; thus, for example, navel orange is more susceptible than Valencia orange (272), and common sour orange is more susceptible than smooth Seville (142).

Under conditions especially favorable for infection, trees in the nursery may be affected by a blight of the foliage, a rot of the buds, and a dieback of the sprouts. Apparently all varieties, including sour orange and Cleopatra mandarin, can be attacked under these conditions (251).

Symptoms

Foot rot lesions usually begin near the bud union. They generally expand more rapidly in an upward direction but can also spread downward into the root crown. First symptoms are water-soaking of the bark in irregular patches of variable size and oozing of varying amounts of gum. In time, the affected bark dries out, settles, cracks longitudinally, and weathers off (Fig. 24). The inner bark develops a brown to sooty discoloration and becomes filled with gum. Gum often accumulates in pockets and forces the bark apart. The cambium is affected and a thin layer of the wood is stained. In advance of the killed tissues, the cambium is slightly yellow and suffused with a clear, watery gum-like substance. Affected tissues impart a fermentative odor.

Callus formation may eventually occur around the healthy margin of the lesion, limiting further progress of the disease. Below ground, delimitation occurs less frequently and girdling often takes place. Secondary organisms may attack the wood and render trees subject to toppling during wind storms.

The effects of foot rot in the tops vary with extensiveness of the trunk lesions. A one-sided attack of the trunk causes a one-sided decline; a girdling lesion leads to a generalized decline. Top symptoms are the same as those produced by any agency that disrupts the flow of nutrients, synthesized food, and water through the vascular system: starvation patterns in the foliage, reduction in leaf and fruit size, leaf drop and dieback, and a reduction in vegetative vigor.

In some cases, lesions heal over temporarily but resume activity at a later date. Trunk lesions rarely

FIG. 24. *FOOT ROT.* Bark lesion in the sweet-orange portion of a trunk. Inclination produced by a hurricane has brought the susceptible portion in contact with the ground, rendering it susceptible to attack by the soil-borne foot rot fungus.

extend higher than 18 in. from the ground. Not all lesions progress until trees are destroyed. Lesions often heal over and trees may show no evidence of having been injured. It has been estimated that in Florida over half the cases of foot rot are self-healing.

Inoculum for infecting the trunks or crowns arises from infected feeder roots, but damage from fibrous root infection in itself is not as conspicuous in Florida as in parts of the world where soils are heavy. The apparent absence in Florida of tree decline due to fibrous root infection is probably attributable to the rapid regeneration of fibrous roots, especially during times when soil moisture is not optimal for fungus development.

Cause

Although a number of different species of *Phytophthora* are known to produce foot rot elsewhere (349), only *Phytophthora parasitica* Dastur (syn. *P. nicotianae* B. de Haan var. *parasitica* [Dastur] Waterh.) has been isolated from foot rot lesions in Florida (70, 143, 363). Both *P. parasitica* and *P. citrophthora* (Sm. & Sm.) Leonian have been isolated from fruits affected by **Brown Rot** (which see). Variations in pathogenicity among different isolates of *P. parasitica* are recognized (143). Incidence of infection is related to zoospore concentration (363).

A simultaneous occurrence of several environmental conditions is required for infection and the development of foot rot.

Conditions that favor fungus growth, multiplication, and infection.—Temperatures between 86 and 90°F are optimum for growth, but *P. parasitica* can tolerate temperatures down to about 50° and up to 99°F. Soil moisture at or near saturation is most favorable for growth, spore production, and movement of zoospores (115). The fungus is very sensitive to moisture fluctuations; its activity ceases when soil dries out. Moisture on the bark is also conducive to infection and to the subsequent development of foot rot lesions (363). A pH of 6.0–7.5 favors the growth and multiplication of the fungus in culture and presumably also in the soil, and the growth rate of the fungus as influenced by pH has been found to parallel the amount of fibrous root infection (115, 331).

Conditions that render the host susceptible.—As mentioned above, citrus varieties differ genetically with respect to susceptibility. Extraneous factors are also important, particularly those that allow entrance of the fungus through wounds and growth cracks (363) and those like splashing rain that carry zoospores into such openings. The low planting of trees and the heaping of soil near the bud union increase the chances of infecting the susceptible scion portion of the trunk.

Factors that disseminate the fungus.—Movement of stock from infected nurseries is thought to be the major way in which the disease is spread from one locality to another (70). Elsewhere in the world, dissemination is known to occur from contaminated flood and irrigation water; in Florida, however, the importance of this vehicle has not been investigated.

Control

A foot-rot-tolerant rootstock serves much as does a masonry foundation to protect a wooden house. If the susceptible scion portion of the tree is too close to the ground, the protection afforded by the rootstock is lost. Trees should be budded high (at least 3 in. above the ground) and when planted in the grove, trees should be set so that the topmost root is just below the ground line. The choice of rootstocks should be governed by information given under *Plants affected*. Adoption of the foot-rot-resistant sour orange should be avoided in areas where the threat of aphid-borne tristeza virus exists, and the use of trifoliate orange and the citranges should be restricted to budding with scion varieties of which exocortis-free budwood is available.

If foot rot lesions develop, they can be treated surgically (70, 96, 272), but this practice is no longer considered economical except for specimen trees. Further expansion of lesions can often be prevented by washing the soil away from the trunks to expose the topmost root and by improving air drainage through the removal of low-hanging limbs and weeds. If lesions encircle more than a third of the trunk, affected trees should be pulled and replaced. Trees that are only partially girdled will often recover (70).

Fumigation of infested replant sites enables new

trees to become established more rapidly (9). Use of methyl bromide is suggested as directed under **Slow Decline.**

Foot rot is easier to avoid than to treat. Grove sites should be well-drained and planted with noninfested, high-budded nursery trees. During cultivation, care should be taken not to heap soil around the trunks and not to cause injury to the bark. Banks should be removed as early as possible after the danger of freeze is over.

To avoid infestation of nursery sites, seeds should be treated in hot water (10 minutes at 125°F) for the control of seed-borne inoculum (29, 171). Infested nursery sites may be decontaminated by pre-plant fumigation described under **Slow Decline.**

FOVEA. A bud-transmissible disease of the Murcott, causing inverse stem pitting and the death of trees.

At one time fovea (Fig. 25A) led to the decline and death of many Murcott trees (182) but with present-day availability and use of registered virus-free clones of Murcott, the disease is no longer prevalent.

Symptoms under the bark may be seen as early as 2 years after budding, but decline of the top does not appear until trees are about 4 years old. The time elapsing between first appearance of decline and the death of trees is approximately 2 years. Removal of bark from an affected tree shows the woody cylinder studded with bristly pegs (Fig. 25B) that fit into holes up to 1 mm in diameter on the inner face of the bark (Fig. 25C). Bristles are clustered at slight depressions along the woody cylinder and are most numerous above the bud union and throughout the trunk but can be found in branches as small as ¾ in. in diameter. The holes in the inner face of the bark resemble the honeycombing symptom of tristeza in sour orange and the inverse pitting occasionally seen in xyloporosis-affected trees, but differ from these two types in being somewhat larger and more variable in size and in not being restricted to areas adjoining the bud union.

The cause of fovea is a bud-transmissible virus. It has been suggested (208, 253) that the symptoms of fovea are produced by the xyloporosis virus and

that they represent a response to infection peculiar to Murcott. This hypothesis is opposed by the facts that, first, Murcotts infected with xyloporosis virus produce the same stem-pitting symptoms (Fig. 25D) commonly found in sweet lime and mandarin hybrids and, second, Clementine mandarin trees on sweet lime that develop fovea in the scion portion of the tree do not develop xyloporosis pitting in the xyloporosis-susceptible sweet lime portion (204).

Fovea can be avoided by propagating with registered virus-free budwood or by planting Murcott seedlings.

FREEZE DAMAGE. See **Cold Damage.**

FRIENDLY FUNGI. See **Entomogenous Fungi.**

FUSARIUM. Genus of the deuteromycetous fungi. *F. lateritium* Nees and *F. solani* (Mart.) Appel & Wr. have been isolated from blighted twigs, and the latter species also from foot rot lesions. *F.* spp. are found frequently in decaying roots. Though Fusarium fungi may be present in many parts of citrus trees (375), they are not considered to be primary pathogens.

GANODERMA. Genus of the basidiomycetous fungi. One species, *G. sessilis* Murrill, has been reported to cause **Wood Rot** (which see).

GAS BURN. Rind breakdowns following ethylene treatments for degreening. *SYN—ethylene burn.*

Several types of injury to the rind develop after fruits have been degreened with ethylene. Oranges and grapefruit may develop scattered brown pits, widespread sunken discolored areas (Fig. 26A), or ring-type blemishes at points where fruits have touched. Though long attributed to ethylene burning, these injuries are now thought to result from the gassing of fruits having rinds weakened by the puddling of dissolved fertilizer dust and of spray chemicals. Occasionally, however, gas burn is encountered in lots of fruit that have been washed before degreening, suggesting that accumulations of chemicals on the rind are not the sole cause of gas burn.

FIG. 25. *FOVEA*. A. Decline of Murcott tree affected by fovea. B. Bristles in trunk of a 7-year-old Murcott tree. C. Fovea in the inner face of the bark. D. Xyloporosis pitting in Murcott as contrasted with symptoms of fovea in B. (Photographs B and C by W. C. Price.)

Rind injuries of similar appearance may be produced by excessive drying of the rind during harvesting and handling. Such injuries are often erroneously referred to as gas burn. They should be recognized as being instances of **Stem-End Rind Breakdown** (which see).

In Temples, similar injuries to the rind (Fig. 26B) may be caused by ethylene alone (139). The amount of damage is related to concentration of the gas and to duration of the exposure. This type of burn may not become apparent until 2–18 days after degreening.

GEOTRICHUM. Genus of the ascomycetous fungi. *G. candidum* Link (syn. *Oospora citri-aurantii* [Ferr.] Sacc. & Syd.) causes **Sour Rot** (which see).

GLASSY SPOT. See Cancroid Spot.

GLOEODES. Genus of the deuteromycetous fungi. *G. pomigena* (Schw.) Colby causes **Sooty Blotch** (which see).

GLOEOSPORIUM. Genus of the deuteromycetous fungi. *G. limetticola* Clausen is the cause of **Lime Anthracnose** (which see).

GLOMERELLA. Genus of the ascomycetous fungi. *G. cingulata* (Stonem.) Spauld. & Schrenk (the perfect stage of *Colletotrichum gloeosporioides* Penz.) causes latent infections in healthy tissue and produces fruiting bodies in debilitated or dead tissue. See **Alternaria Leafspot, Anthracnose, Symptomless Infections.**

GOLD THREAD. See Dodder.

GRANULATION. A drying out of the flesh of fruits on the tree. *SYN.*—crystallization, ricing.

The flesh of fruits on the tree may become dry either because of **Cold Damage** (which see) or because of a condition known as granulation. In tree-frozen fruit, juice sacs collapse, shrivel, and separate from each other and from the segment walls; in granulated fruit, conversely, sacs do not separate but become filled with yellowish or grayish-white solid matter. Granulation does not affect the firmness of the fruit but it does reduce the weight. Granulated vesicles are tasteless because of low sugar and acid contents (297). The drying-out process begins at the stem end and may progress until the entire flesh is involved, but granulation does not increase after fruits are harvested (18). Navel oranges grown in Florida are particularly subject to granulation.

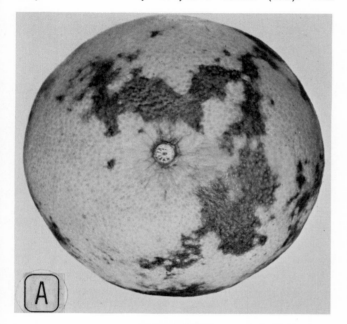

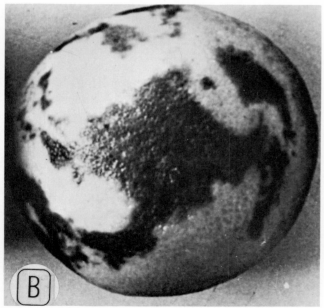

FIG. 26. *GAS BURN.* A. Peel injury after ethylene degreening of unwashed sweet-orange fruit. B. Gas burn in washed Temple fruit. (Photographs by W. Grierson.)

The causes of granulation are complex and appear to involve rootstock influences, genetic factors, maturation processes, and soil-moisture relations (18).

To some extent granulation may be reduced by supplemental irrigation, by applications of 2,4-D, and by early harvesting (88).

GRAPEFRUIT SCAB. See **Scab.**

GRAPHIS. Genus of the Lichens. *G. afzelii* Ach. occasionally produces white, lenticular growths on sound wood below twigs that have died back. See **Lichens.**

GRAY FUNGUS. *Podonectria coccicola* (Ell. & Ev.) Petch, one of the so-called "friendly fungi." See **Entomogenous Fungi.**

GREASY MELANOSE. See **Greasy Spot.**

GREASY SPOT. A spotting of the leaves that leads to defoliation; also a stippling of the rind in sweet-orange fruits. SYN.—greasy melanose, black melanose.

The earliest account of diseases affecting citrus in Florida, written in 1896, made mention of what is known today as greasy spot (323). Subsequent writers recognized the disease to be of widespread occurrence but considered it to be of little importance (272, 305). It was not until the 1950s, when oil sprays were replaced by organic scalicides, that greasy spot became a serious problem.

The economic consequence of greasy spot is defoliation. Incidence of infection and amount of leaf drop vary from year to year. Instances have been known of groves in which 80 per cent of the foliage dropped because of this disease. Since 1956, greasy spot has been included in annual revisions of the *Florida Citrus Spray and Dust Schedule* (4) as a problem requiring routine preventive measures.

Plants affected

Greasy spot may be found in the foliage of all varieties of citrus grown in the state. Spotting has been said to be most severe in grapefruit and lemon, but that these varieties defoliate more

readily than others is not yet an authenticated fact.

Symptoms

The earliest symptom visible to the unaided eye consists of slight blisters, predominating on the undersurfaces of leaves. On upper surfaces, each spot is marked by a chlorotic dot somewhat larger in size than the edema. In time, blistered areas become necrotic and range in color from an initial light orange-brown to an ultimate chestnut-brown or black. As generally seen in the grove, spots resemble irregular, diffusely margined flecks of dirty grease (Fig. 27A). Lesions on the undersurfaces eventually penetrate the leaf and produce the same greasy appearance on the upper surface. Affected areas are usually scattered at random over the leaf but at times may be restricted to the portion of the blade on one side of the midrib. Leaf margins are often bordered on the undersurface by a continuous band of stained tissue (Fig. 27B) and on the upper surface by a corresponding band of chlorotic tissue (Fig. 27C).

Infections occur mostly in summer, and spots appear 2–8 months later, depending on the host variety. The interval from first symptoms to ultimate leaf drop is extremely variable, though severe defoliation frequently occurs by the end of winter. In some varieties of lemon, infected leaves may drop without the development of conspicuous necrotic lesions.

Greasy spot has long been considered a disease confined to the foliage, but recent investigations show that the causal organism may also attack the fruits, causing a fine stippling (see title page) that interferes with coloring (362). The stippling takes the form of pinpoint black specks that become discernible only after colorbreak. With magnification, each speck is seen to consist of a necrosis of the stomata and of a few underlying cells. Fruits may be affected to such an extent that they are dropped a grade in the packinghouse. There is no increase in the size of specks with time after harvesting.

Cause

For many years, greasy spot was believed to result from the feeding of rust mites (21, 328). Not until 1952 did Japanese workers identify the cause as a Cercospora-type fungus with *Mycosphaerella horii* Hara as the perfect stage (324). A subsequent search for the causal organism in Florida led to a report that in Florida, too, a Cercospora fungus was involved (105). The fungus was given the name *C. citri-grisea* Fisher (106). More recently, a Mycosphaerella fungus encountered sporulating abundantly in decomposed fallen citrus leaves was found to be the causal agent (360). This Mycosphaerella differs in several respects from the one in Japan. Spores of the imperfect stage are rough-walled, 0–9-septate, cylindrical, and 6–50 x 2.0–3.5 microns in size. They are produced sparsely on

simple conidiophores that arise from rough-walled hyphae on the surface of leaves on the tree, and they belong to the genus *Stenella*. The perfect stage is *Mycosphaerella citri* Whiteside (364). Black-walled perithecia occur subepidermally in densely packed groups on decomposed leaves. Ascospores are hyaline, straight or slightly curved, and mostly 8.5 x 2.5 microns in size.

The inoculum for infection comes mainly from the perithecia. Ascospores are released during rainfall. The time of maximum spore discharge varies from year to year and from grove to grove, depending on the occurrence of rainfall; in general, however, greatest spore discharge occurs during the earlier part of the rainy season. Frequent or prolonged rainfall results in the exhaustion of ascospore supply by accelerating leaf decomposition. Unless there is subsequent leaf drop to replenish the supply of ascospores, the inoculum level will be reduced as the rainy season progresses. While most infection occurs in June or July, it may take place any time when there is sufficient rainfall.

Control

Infection may take place throughout the year but because most infections occur during the earlier part of the summer, one spray applied in June or July usually affords satisfactory control. Materials currently recommended are copper (1.25–2.50 lb. metallic/500 gal. water), zineb (5 lb. of 75 per cent wettable powder/500 gal. water), and oil (1 per cent FC 435-66). Some control of the disease in the spring flush is provided by the postbloom copper application intended for the control of melanose.

A reduction in the amount of inoculum might be expected from disking under the leaf litter that gives rise to perithecia.

GREASY SPOT RIND BLOTCH. See footnote to **Pink Pitting.**

GREENING-LIKE DISEASE. See **Sandhill Decline, Young-Tree Decline,** and **Blight.**

GREEN MOLD. One of the postharvest rots of citrus fruits. *SYN.*—pinhole rot, blister rot.

Green mold (caused by *Penicillium digitatum*

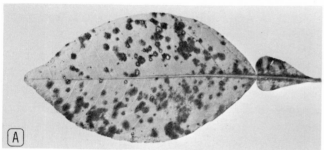

FIG. 27. *GREASY SPOT.* A. Undersurface of grapefruit leaf, showing oily, brown swellings. B. Undersurface of lemon leaf, showing confluent type of greasy spot. C. Upper surface of leaf in B, showing chlorosis produced by underlying necrotic lesions.

Sacc.) and blue mold (caused by *P. italicum* Wehmer) are the most frequently encountered rots of citrus fruits. Of the two, green mold (Fig. 28) is by far the more common in Florida. While both are mainly postharvest problems, they occasionally affect fruits on the tree. At times, both fungi occur on the same fruit.

Green mold is characterized by the greenish color of the powdery growth, by the wrinkled surface of the colony, by sporulation on the exterior of the rind, by a broad border of white mycelial growth, and by an indefinite ring of watersoaked rind around the colony.

FIG. 28. *GREEN MOLD.* Sporulation of fungus at left, white mycelial border at center, and faint water-soaked band in advance of white border.

Both green and blue mold first appear as soft water-soaked areas up to ½ in. in diameter (pinhole-rot stage), enlarging within 36 hours to a diameter of several inches (blister-rot stage). Soon after enlargement, sporulation takes place over the water-soaked areas.

Spores of *P. digitatum* as well as those of *P. italicum* are universally present in the atmosphere. They enter fruits through breaks or weaknesses in the rind. In contrast to blue mold, green mold does not pass from infected fruits to adjacent sound fruits by contact.

Sanitary practices should be followed to hold down the spore load, and careful handling of fruit should be practiced to minimize injuries to the rind. Green and blue mold can be controlled by postharvest applications of such fungicides as biphenyl, sodium o-phenylphenate (SOPP), and thiabenzidole (TBZ) (220).

GREEN SCURF. See **Algal Disease.**

GREEN SPOT. See **Oleocellosis.**

GUM DISEASE. See **Rio Grande Gummosis.**

GUMMOSIS. See **Rio Grande Gummosis.**

HAIL DAMAGE. Injuries to fruit and bark from hailstorms.

Symptoms of hail damage vary with the size of hailstones. Small stones puncture oil glands, causing escaped oil to burn the rind in the manner of oleocellosis. Larger stones may pock the rind and leave dark, scabby blemishes as the fruits expand (Fig. 29A). Golf-ball-sized stones may break the rind to expose the flesh (Fig. 29B), may knock mature fruit off the tree, and may tear the bark off branches (Fig. 29C).

HAPLOSPORELLA. Genus of the ascomycetous fungi. One species has been equated with *Sphaeropsis tumefaciens* Hedges, the cause of **Sphaeropsis Knot** (which see).

HEART ROT. See **Wood Rot.**

HELLBIND. See **Dodder.**

HEMICYCLIOPHORA. Genus of the phytophagous nematodes. *H. arenaria* Raski (one of the sheath nematodes) has been shown to be pathogenic to citrus, but its economic consequences remain to be evaluated. See **Nematodes.**

HIRSUTELLA. Genus of the basidiomycetous fungi. *H. besseyi* Fisher is parasitic on Florida red and purple scale, and *H. thompsonii* Fisher on citrus rust mite. See **Entomogenous Fungi.**

INFECTIOUS VARIEGATION. A mottling and at

times a distortion of leaves caused by a graft-transmissible virus; a disease of minor importance in Florida.

Trees of sweet orange, grapefruit, and rough lemon are encountered occasionally that show a conspicuous light-green or whitish blotching of mature leaves (Fig. 30). Affected foliage is usually restricted to a single branch or twig. Other symptoms may include a reduction in leaf size and a crinkling of the blade (not to be confused with **Crinkle-Scurf,** which see).

When infected buds are grafted into sour orange or Eureka lemon indicator seedlings, mottling develops in the seedling foliage in 20–90 days,

sometimes preceded by the development of a psorosis-like vein flecking in young leaves. Some of the subsequently formed young leaves of lemon seedlings may show a pinpoint stippling and crinkling that persist after leaves have matured. In the field, affected trees do not appear to suffer reduction in vegetative vigor or fruitfulness.

Cause of the variegation and occasional crinkling is a virus that appears to be closely related to the one causing **Crinkly Leaf** (which see). Both viruses are sometimes assigned to the psorosis virus group.

The infectious variegation virus is transmitted from buds of infected trees to healthy nursery plants by grafting. The partially purified virus has

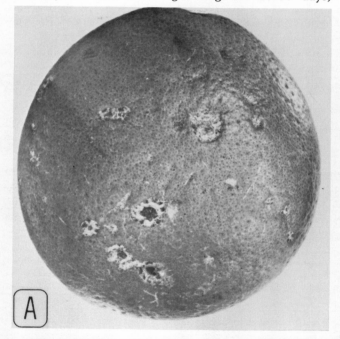

FIG. 29. *HAIL DAMAGE.* A. Development of scabby blemishes on sweet orange 3 months after a hailstorm. B. Sweet oranges ruptured by golf-ball-sized hailstones. C. Rupture of the bark of twigs by large hailstones.

been transmitted mechanically to citrus and to *Phaseolus vulgaris, Dolichos lablab, Tithonia speciosa, Crotalaria spectabilis,* and *Vigna sinensis* (135, 224).

The disease can be avoided by selecting bud-wood from noninfected trees for propagation. Since there are several instances that suggest transmission of the virus through seed (173, 224), control would also require the avoidance of seed source trees with symptoms of infectious variegation. Under experimental conditions, the virus has been inacti-

FIG. 30. *INFECTIOUS VARIEGATION.* Leaves of Eureka lemon showing some of the puckering and mottling patterns produced by citrus infectious variegation virus. (Photograph by J. M. Wallace.)

vated by holding infected plants for 4 weeks in a heat chamber at a temperature of 38°C (224).

INSPISSOSIS. See Nematospora.

INTERNAL DECLINE. See Endoxerosis.

JUVENILE SPOT. A minor disorder of young grape-fruit trees characterized by spotting of leaves and defoliation.

Juvenile spot is an abnormality found occasionally on grapefruit foliage. It may also occur rarely in orange and tangerine trees. The name derives from the occurrence of the disorder in trees less than 6 years old.

Though juvenile spot may defoliate young grapefruit trees, it occurs with such infrequency as to be of little importance. At worst, spots are sometimes mistaken for those of bacterial canker.

In addition to Florida, juvenile spot has been seen in Argentina, Syria, South Africa, and Japan.

Symptoms appear on whichever side of the leaf happens to be turned upward. Spots consist of round, chocolate-colored, gummy platelets 1/32–3/8 in. in diameter, emarginated by yellow halos (Fig. 31). When young, each spot contains a central mound of gum; later, spots become crateriform, giving the appearance of having been punctured. Sometimes platelets are concentrically ringed, suggestive of bacterial canker lesions.

The cause of juvenile spot is not known though bacteria, fungi, and fertilizer burn have for the

FIG. 31. *JUVENILE SPOT.* Symptoms on leaf of grapefruit.

most part been eliminated as possibilities (197). The pattern of distribution in groves indicates the trouble to be noninfectious.

KNOT DISEASES. See **Brevipalpus Gall, Sphaeropsis Knot, Wart.**

LAMINATE SHELLING. An apparently noninfectious eruption of the bark in the rootstock portions of trees on trifoliate orange.

Laminate shelling, though of infrequent occurrence, deserves mention because of the need to distinguish it from exocortis scaling. Both bark troubles are similar in appearance and in variety affected. Laminate shelling, however, is characterized by patches consisting of tightly packed beadlike projections, ¼ in. in elevation, that partly circle the bark in trifoliate-orange portions of the trunk. The patches occur in arcs ½–5 in. long and ¼–2 in. wide. The many-layered structure of the projections resembles the type of bark found in citrus relatives such as *Hesperethusa* and *Zanthoxylum*. Affected trees are not usually stunted, even at 30 years of age, and this further suggests that laminate shelling results from a genetic abnormality rather than from the virus of exocortis (209).

LATENT INFECTIONS See **Symptomless Infections.**

LEAF BLOTCH. See **Lime Blotch.**

LEMON SCAB. See **Scab.**

LEPRA EXPLOSIVA. See **Leprosis.**

LEPROSIS. A mite-associated cankering of fruits, leaves, twigs, and branches. *SYN.*—Florida scaly bark, lepra explosiva, nailhead rust.

At one time, leprosis was known throughout the citrus-growing area of Florida, but today it verges on extinction. In parts of South America, however, leprosis remains a serious problem.

Plants affected

Though the causal agent may be present on all varieties of citrus (201), it produces economic damage only in early and midseason varieties of sweet oranges. Sour orange is also susceptible. A few lesions may occasionally be found in trees of mandarin, grapefruit, and rough lemon.

History and distribution

From its focal point in Pinellas County, where it was first observed in the 1860s, leprosis spread eastward across the state. In attempts to stem its advances, the Florida State Plant Board prohibited the movement of citrus trees and parts thereof from 120 affected locations in 9 counties. Despite this measure, leprosis continued to spread, and by 1925, 17 of the state's citrus-growing counties had been invaded. Then suddenly in the late 1920s, leprosis began to disappear. Today the disease is known only in a few poorly managed groves in Volusia and Sumter counties (201).

Importance

The destructive capabilities of leprosis are awesome. Thousands of acres of citrus have been attacked and rendered unproductive in Argentina, Uruguay, Paraguay, and Brazil (212).

Fruits are spotted and may drop prematurely. Affected fruits that are harvested must be culled in the packinghouse. At the height of the epidemic in Pinellas County, 20,000–30,000 boxes of oranges were rejected annually, and in badly affected groves, from 35 to 75 per cent of the crop dropped before maturity.

Lesions similar to those on fruit develop on young, green shoots. At first, lesions are barely visible, but with time they enlarge until they may girdle the trunk or limb that develops from the attacked shoot. Girdling that results from a few lesions does not usually lead to dieback, though terminal growth and cropping are reduced.

Leaves also develop spots and shed prematurely, but defoliation is negligible.

Symptoms

On fruits, scattered chlorotic spots become visible while the rind is still green. With time, spots enlarge to as much as ¼ in. in diameter and turn rusty brown (Fig. 32A), hence the old term "nailhead rust." Centers are usually sunken and sometimes superficially cracked. Before fruits break color, the necrotic spots are surrounded by yellow halos. Lesions do not penetrate the flesh.

On leaves, spots resemble those on fruits except that they are somewhat angular (Fig. 32C). Each gummy plaque with its yellow halo is visible on both sides of the leaf. Spots occur predominantly along the leaf margins. There are seldom more than 10 spots per leaf, and there are rarely more than a few leaves affected.

Bark scaling (Fig. 32E) is the most dependable symptom of leprosis, because twig and branch lesions persist after affected fruit and leaves have dropped. First symptoms are visible several months after shoots emerge from the bud. As on fruits, lesions begin as pinpoint yellow spots. With hardening of the bark they enlarge, turn brown, and develop thick, resinous scales. Old leprosis bark cankers can still be found on trunks and in limbs in treetops that were attacked many years previously (Fig. 33A).

The symptoms of leprosis in Florida (Fig. 32A, C) differ somewhat from those of leprosis in South America (Fig. 32B, D) (212).

Leprosis scaling is sometimes confused with psorosis scaling. In leprosis, the scales are reddish brown, resinous, and much thicker than bark, whereas in psorosis, scales are bark-colored, sometimes streaked with gum, and are as thin as normal bark. Differentiation is more difficult when scales have been removed. In such cases, however, leprosis is indicated if the lesion is swollen, whereas psorosis is involved if the lesion is slightly sunken.

Leprosis has at times been confused also with **False Leprosis** (which see).

Cause

In 1911, the cause of leprosis was reported to be a fungus, *Cladosporium herbarum* var. *citricolum* Farl. (90). This attribution persisted until 1940 when Argentine investigations showed the real agent to be a false spider mite, *Tenuipalpus pseudocuneatus* Blanchard (119). In Florida the species associated with leprosis is the closely related *Brevipalpus californicus* (Banks) (Fig. 33B) (177). Another false spider mite, *B. phoenicis* (Geijskes), is common on Florida citrus, being exceeded in frequency only by the rust mite, *Phyllocoptruta oleivora*, but it is incapable of producing leprosis (196). *B. phoenicis*, however, is capable of producing a galling and killing of citrus seedlings in the nursery (see **Brevipalpus Gall**). Brevipalpids are often overlooked because they are five times smaller than the citrus red mite *Metatetranychus citri* (McG.).

The relation of mites to leprosis was demonstrated by infesting sweet orange seedlings with *B. californicus* under caged conditions (196). Further evidence of the role that mites play as the inciting agent was obtained from spray trials showing that leprosis can be controlled with acaricides (211).

Lesions can be reproduced by infesting seedlings with nymphs hatched in the laboratory from eggs and by grafting seedlings with leprosis-affected bark, but neither line of evidence indicates with certainty whether leprosis is due to a mite-injected toxin or to a mite-borne localized virus (196).[1]

Control

Excellent control of both Brevipalpus mites and leprosis can be obtained with an annual postbloom application of either wettable sulfur, 10 lb./100 gal. water, or chlorobenzilate, ¼ pint 45.5 per cent liquid/100 gal. water (163, 211).

The causal mite is still present in widely scattered localities of Florida (201); therefore, leprosis could again become a problem if indicated miticides were omitted for several years from the spray program (190).

LEPTOTHYRIUM. Genus of the deuteromycetous fungi. *L. pomi* (Mont. & Fr.) Sacc. is the cause of **Flyspeck** (which see).

LICHENS. A diverse group of plants consisting of fungi and algae intertwined to form classified species; those on citrus bark form nonpathogenic, moss-like patches.

To the botanist, lichens are of interest because they exemplify a curious union of an alga and a

1. Subsequently, Kitajima, Muller, Costa, and Yuki (Virology 50:254–258. 1972) have reported short, Rhabdovirus-like particles to be associated with leprosis in Brazil.

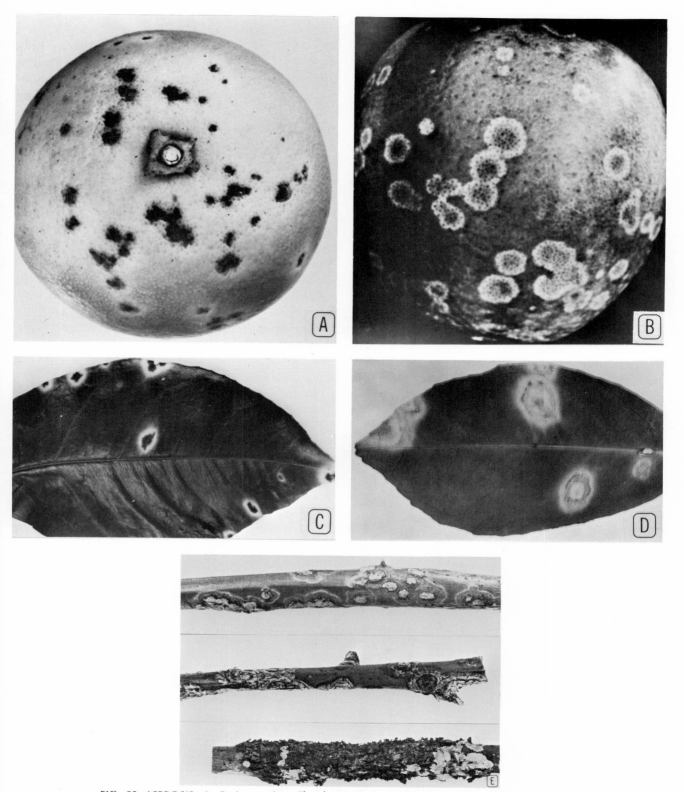

FIG. 32. *LEPROSIS.* A. Fruit spotting, Florida. B. Fruit spotting, Venezuela. C. Leaf spotting, Florida. D. Leaf spotting, Venezuela. E. Twig lesions, Florida. Top: Year-old lesions. Middle: Two-year-old lesions. Bottom: Confluent lesions at advanced stage of scaling. (Photographs B and D by M. Hruskovec.)

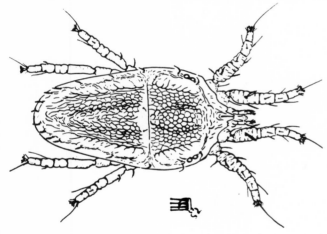

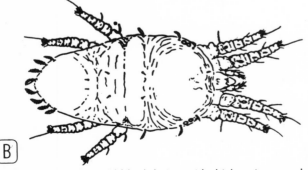

FIG. 33. *LEPROSIS*. A. Old bark lesion with thick resinous scales on trunk of sweet orange—the result of an infestation of *Brevipalpus californicus* some 70 years earlier. B. *Brevipalpus californicus* (Banks), the inciting agent of leprosis. Top: Female, dorsal aspect with enlargement of claw. Bottom: Nymph, dorsal aspect. (After Prichard and Baker.)

fungus to survive conditions under which separately both might perish. The fungus provides the alga with protection from desiccation, and the alga furnishes the fungus, as well as itself, with photosynthesized food. The ascomycetous and basidiomycetous fungi involved have lived so long in association with their green and blue-green algal mates that the two components are rarely found separate in nature. In fact, botanists consider the 15,000 different lichens as distinct species, indefensible though this convenience may be from a strict taxonomic point of view.

To the experienced citrus grower, lichens are of no interest whatever, but for the newcomer to Florida, they are often a source of concern. Short of interfering slightly with the host's respiration and transpiration and of harboring some noxious mites and insects, lichens are harmless to citrus. Their presence, however, does betray a certain deficiency in grove care since lichens indicate that infested trees have not been sprayed regularly with copper or oil for the control of diseases and pests.

If these multicolored, paper-thin, irregularly lobed incrustations on the trunks and limbs (Fig. 34A, B) appear offensive, or if heavy infestations are present, they may be removed by sprays containing neutral copper (3.75 lb. metallic/500 gal. water) or 1.3 per cent oil emulsion. Best results are obtained by spraying lichens while they are dry.

Occasionally one species, *Graphis afzelii* Ach., is seen on sound wood below twigs that have died back (Fig. 34C), suggesting, though erroneously, that this organism is the cause of the dieback.

LIGHTNING DAMAGE. Injuries to trees from lightning, either by direct strikes or by ground discharges.

Although lightning may injure and even destroy citrus trees, its effects are more curious than pecuniary. The sudden decline of a few trees struck by lightning often elicits more interest than do some gradual declines of real importance.

Lightning discharging in a grove affects a few to several dozen trees. Usually 1 or 2 trees within an area are severely damaged or killed while nearby trees show symptoms varying from wilting to bark scalding. Lightning does not rip open the bark of citrus trees as is the case with some forest trees. A direct hit to the trunk does little more than kill a narrow strip of bark, ½–2 in. in width, down to the surface of the soil (306). At the base of the trunk, the strip may spread out to involve the entire crown. Nearby trees may show a dieback of some twigs in the canopy and a characteristic killing of the superficial layers of the bark between the nodes of twigs. The bark at the base of petioles and thorns remains green (Fig. 35). At first, the affected bark is greenish-yellow; later, it becomes brownish, heightening the contrast between damaged bark and the green islands or bands of intact tissue. Where the cambium is not affected, it forms new tissue which eventually elevates the injured areas, leaving slightly raised, roughened plaques.

Lightning travels a course of least resistance. If rains have wet the bark, the charge moves along the bark surface, causing damage only to the twigs. Moisture-deficient trees suffer more than those well supplied with water. Deep-rooted trees are damaged more than shallow-rooted ones, and trees on clayey soils are more resistant than those on sandy soils. The severity of damage diminishes with distance from the center of the strike. Trees only slightly affected survive; those in which there is extensive killing of roots and limbs and a subsequent invasion of the weakened bark by secondary organisms seldom recover.

LIGNOCORTOSIS. See **Lime Blotch.**

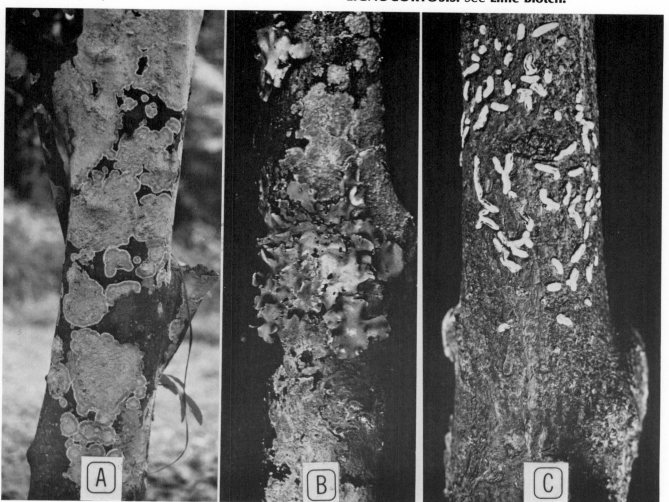

FIG. 34. *LICHENS.* A., B. Several types of lichens encountered on the bark of citrus. C. *Graphis afzelii* Ach., a lichen occasionally found on bark below twigs that have died back. None of the lichens is harmful to citrus.

LIME ANTHRACNOSE.

LIME ANTHRACNOSE. A bud-blasting, fruit-spotting, shoot-killing disease of limes, particularly of the variety Key (also called Florida Common, West Indian, and Mexican). *SYN.*—withertip, blossom blight.

The cultivation of Key limes has never been important in Florida because of the variety's intolerance to cold and lime anthracnose. Nevertheless, from time to time efforts are made to grow this lime on a commercial scale because of the lucrative returns from peel oil.

Lime anthracnose has been a limiting factor in the growing of seedy acid limes in humid areas. In the 1920s, the disease drove the lime industry out of many islands of the Caribbean (217). Even in drier areas of the West Indies where limes are still grown, the major cost of production is the constant spraying required to protect new growth.

Plants affected

There are no reports of the lime anthracnose fungus attacking any variety of citrus other than the small acid lime *(Citrus aurantifolia)*. Of 56 com-

FIG. 35. *LIGHTNING DAMAGE.* Note at base of petioles the islands of healthy, intact bark that stand out against scalded portions of the twig.

mercial and exotic varieties of citrus inoculated with the causal organism in the field and in the greenhouse, only the West Indian and the Dominican Thornless (a sport of the West Indian) showed susceptibility. The large seedless lime known as the Tahiti or Persian is immune (120).

Other varieties of citrus such as rough lemon, table lemon, sweet orange, and grapefruit have problems referred to as anthracnose, but the fungus involved is different from the one causing lime anthracnose. Diseases attributed to *Colletotrichum gloeosporioides* are discussed under **Anthracnose.**

Symptoms

The lime anthracnose fungus affects all tender tissues of the host including buds, leaves, fruits, and shoots. Repeated and uncontrolled attacks reduce canopies and lead ultimately to tree decline.

The most serious consequence of infection is a reduction in cropping due to blasting of buds and blossoms. Attacked buds fail to open, turn brown, and drop. Buds that flower may subsequently be attacked and destroyed.

Young leaves develop a marginal necrosis that later leads to leaf distortion. Lesions that form within the leaf blade ultimately produce shothole effects. The acervuli or fruiting structures of the fungus can be seen in dead tissue; they appear as bright, flesh-colored points, later turning rusty brown. Severe attacks on the foliage diminish the photosynthetic potential of the canopy.

Young shoots are attacked and killed back for several inches from the tips. Occasionally, shoots are girdled some distance down from the terminals, causing the green tip portion to topple over.

Fruits attacked by lime anthracnose usually shed prematurely. Those that remain attached to the tree develop round, necrotic lesions (Fig. 36) that may be confused with those of canker or scab. A lime anthracnose lesion as examined under a hand lens is characterized by a brownish corky wart that at the base is surrounded by a narrow ring of scar tissue. In contrast to canker, there is no conspicuous water-soaked halo around the lesion, and the warty growth is found only on the fruit and not, as in canker and scab, on the leaves and twigs. Confusion with scab can further be avoided if it is remembered that scab does not attack West Indian lime. Lesions produced by lime anthracnose often lead to fruit distortion and peel splitting.

Cause

Lime anthracnose results from the infection of lime tissue by the fungus *Gloeosporium limetticola* Clausen. The older literature attributed the disease to a similar fungus, *Colletotrichum gloeosporioides,* but later investigations showed the two to be distinct on the basis of pathogenicity, morphology, and behavior in culture (66, 235).

Rain or dripping dew is required for spore dissemination and germination. Only immature tissues

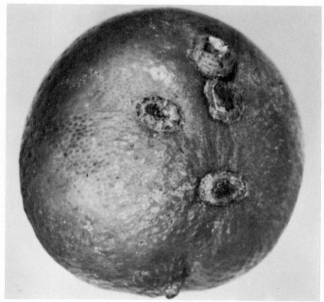

FIG. 36. *LIME ANTHRACNOSE.* Lesions on the rind and consequent distortion of the fruit. (Photograph by M. Hruskovec.)

can be infected. Once leaves harden and develop waxy cuticles and high concentrations of certain cell components (possibly isopimpinellen), they become immune (225). Fruits are no longer susceptible after their equatorial diameters exceed ¾ in.

The fungus is thought to live from one season to the next in dead twigs where it forms acervuli that release spores during later periods of rain.

Control

Although copper-containing sprays are known to be lethal to the causal organism, their applications are difficult during the prolonged rainy seasons of the tropics. Further difficulties are encountered because of the everflushing nature of the host.

Despite applications every 5 days during infection periods, the degree of control is often disappointing.

The long-recognized need is for varieties of the West Indian lime that are tolerant or immune. Some selections tested under Florida conditions have shown promise, but the oil from fruits of these trees has failed to meet the rigid standards set by the essential oils industry (173).

LIME BARK DISEASE. A basal bark rot of Tahiti lime trees.

Trunks and the larger limbs of Tahiti lime trees are often affected in Florida by a bark and wood rot (329, 330). The lesions resemble those caused by foot rot except that they generally occur some distance from the bud union. Limbs above a lesion eventually die. When the trunk is girdled completely, the entire tree is killed.

At one time, lime bark disease was thought to result from infection by the fungus *Diplodia natalensis.* Today, however, it is considered that this fungus invades tissues only after they have been weakened by other agencies. In one experimental planting that was kept under observation for 5 years, the disease continued to show a random distribution and this led to the suggestion that lime bark disease is the result of a genetic abnormality (73). In Brazil, the disease is attributed to the virus of exocortis (288).

LIME BLOTCH. A destructive disorder in certain clones of Tahiti limes and lemons (?) causing branch lesions, sectored fruits, and a chimera-like chlorosis in foliage. *SYN.*—blotch, leaf blotch, lignocortosis, wood pocket.

Many seedless lime plantings in Florida have been abandoned because of the depredations of lime blotch.

The disorder closely resembles wood pocket, a trouble in California that affects certain clones of semidense Lisbon lemons (37). Whether a genetic abnormality in one species can be regarded as synonymous with that occurring in another species is questionable, but the symptomatology and control of blotch and wood pocket are so much alike that from a practical standpoint, both disorders can be considered under the same heading.

The most conspicuous symptom of lime blotch is a variegation of mature leaves (Fig. 37A). Chlorotic areas range in size from streaks to sectors one-half the area of the leaf. Blotching is confined to one side or the other of the midrib, with the midrib usually forming one of the borders. The number of affected leaves on a tree varies from year to year, but no more than a small amount of the foliage is ever involved.

Associated with leaf variegation are occasional fruits showing sectorial–chimera-like, slightly sunken, olive-brown bands of rind extending from stem to stylar end (Fig. 37C). The flesh is not affected, but in storage sectored fruits are particularly prone to decay.

Twig, limb, and trunk symptoms consist of sunken areas beneath the bark (Fig. 37B). Eventually the bark ruptures, opening the way for weakly pathogenic fungi such as *Diplodia* and *Phomopsis*. Cross sections of affected limbs show necrotic sectors in the wood under the sunken bark. At advanced stages of the disorder, trees decline to the point of worthlessness.

The incidence of blotch in Tahiti lime trees varies from grove to grove. A survey of 21 plantings in Dade County showed an average of 27 per cent of trees to be affected (128), and a survey of 12 plantings in the Ridge area showed incidence to range from 4 to 58 per cent (198). The apparently related wood pocket disease has been seen only occasionally in Florida lemon trees.

At one time blotch and wood pocket were considered to be virus diseases, but recent attempts to transmit them by budding have failed to implicate a virus (37, 73, 198). Results did demonstrate, however, that symptoms may reappear in sprouts of buds taken from affected branches, and that symptoms may be transmitted through some of the seeds of affected fruits. Such evidence has led to the assumption that blotch and wood pocket result from genetic aberration.

For the control of blotch, considerable success has been obtained from a careful selection of budwood for propagation. Selections made by the Florida Budwood Registration Program from apparently blotch-free trees have resulted in thriving plantings that are still symptomless after 10 years.

LOVE VINE. See Dodder.

LUMPY RIND. Swollen areas of the rind underlain by pockets of gum in the albedo. *SYN.*—boron deficiency.

Grapefruit, and to a lesser extent oranges, occasionally develop hard, raised areas covered by intact rind. A pocket of gum in the albedo underlies each lump. The brown pockets are surrounded by water-soaked tissue. Lumpy rind is found particularly in the smaller fruits of declining trees showing zinc chlorosis (272). Symptoms resemble those of impietratura, a virus disease not known to occur in Florida (189). Lumpy rind is thought to result from boron deficiency. Similar symptoms have been seen in trees sprayed with arsenic.

MANGIATO D'AGRO. See Rumple.

MEASLES. A minor disorder of the foliage of certain citrus varieties; characterized by closely spaced, pinhead-sized, slightly raised, brownish spots. *SYN.*—big measles, pinhead yellow spot, speckle.

Trees are occasionally seen in Florida with a stippling in mature leaves (Fig. 38A). Individual spots measure approximately 1/16 in. in diameter and are separated from each other by an equal distance. On upper surfaces of leaves, spots are chlorotic; on undersurfaces, dirty white to brown, and slightly raised. Affected leaves are generally restricted to certain limbs. Measles has been found most frequently in varieties of sweet orange, less often in those of grapefruit, sour orange, and Tahiti lime. The condition has also been seen in seedlings of Washington navel orange and Orlando tangelo grown in vermiculite.

"Big measles" (Fig. 38B) appears to be a variant type. Spots are larger, about 3/16 in. in diameter, and more widely separated. This abnormality has been seen in budded Valencia sweet orange trees and in seedling navel orange trees.

The cause of measles remains unknown though microorganisms, viruses, and minor element defi-

FIG. 37. *LIME BLOTCH.* A. Leaf symptoms on Tahiti lime. Chlorotic areas are buff to yellowish-green in color and conspicuous under reflected light. B. Longitudinal bark splitting in Tahiti lime twigs. C. Olive-brown sectoring in Tahiti lime fruits.

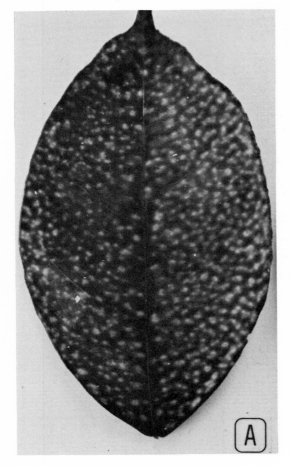

FIG. 38. *MEASLES*. A. Symptoms on sweet-orange leaf. B. Big measles, apparently a variant type, on sweet-orange leaf.

ciencies have for the most part been eliminated as possible causes (197). In some respects, big measles resembles yellow spot, but sodium molybdate sprays have failed to correct the disorder.

The scattered distribution of trees affected by measles and big measles suggests that the troubles are noninfectious.

MELANOSE. A fungus disease forming pinhead-sized corky pustules on young leaves, twigs, and fruit.

The economic consequence of melanose is rind blemishing that reduces the grade of fruit marketed fresh. At one time, the disease was a major problem in Florida, but its importance has diminished now that 80 per cent of the state's fruit goes to market in processed form. For the fresh-fruit grower, however, melanose remains a serious problem.

In addition to melanose, the causal fungus also produces an important decay in harvested fruit (see **Phomopsis Stem-End Rot**).

Plants affected

All varieties of citrus are affected by melanose to some extent. Grapefruit appears more susceptible than orange.

Symptoms

Melanose infections are readily recognized by the reddish-brown to black, pinhead-sized pustules that appear 1–2 weeks after infection on young leaves, stems, and fruits. Pustules occur either as separate dots or as confluent masses that impart a varnish-like, buff or brown coating over affected parts. The coating resembles the russeting caused by feeding injuries of the citrus rust mite, but can be distinguished from it by the sandpapery feel imparted by the pustules. Under the hand lens, pustules appear as dome-shaped eruptions varying in size from pinpoints to pinheads, with bases surrounded by buff-colored halos of scar tissue which in turn are outlined by brown rings (see title page).

In leaves, translucent dots (Fig. 39A) appear 4–7 days after penetration by the causal fungus. Epidermal and subepidermal cells (up to 6 cell layers deep) undergo a gummose degeneration. Intercellular spaces become filled with a water-insoluble, gum-like material. A slight water-soaked depression (marking the site of the eventual pustule) forms on the surface as a result of the dissolution and collapse of affected cells. About a week after inoculation, a corky layer develops underneath the disorganized tissue. With expansion of this layer, the cuticle ruptures to release the pent-up gum which on exposure to the air turns brown and hardens (6). At this stage, the melanose pustule is fully formed (Fig. 39B). Pustules may occur on either surface. Occasionally, single pustules cause the leaf to develop conical pimples containing within their craters the melanose lesions—symptoms that may be confused with

similar structures formed by the scab fungus. Leaves with many pustules become malformed (Fig. 41B), lose their green color, and drop prematurely.

On fruits, the development and appearance of individual pustules are similar to those on leaves (Fig. 40A). There is, however, a much greater tendency than on leaves for pustules to converge and to produce solid or tear-streaked areas of scar tissue (Fig. 40B). As the fruit enlarges, the solid patches crack, producing roughened conditions known as sharkskin and mudcake melanose. Fruit severely infected while very young may be dwarfed and shed prematurely (93).

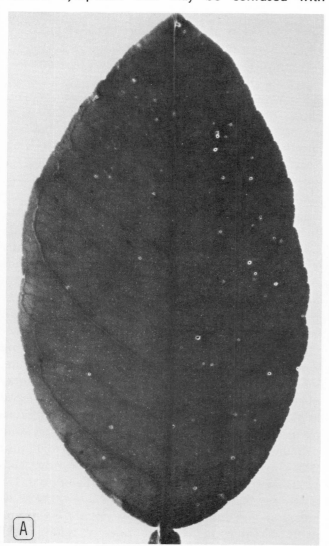 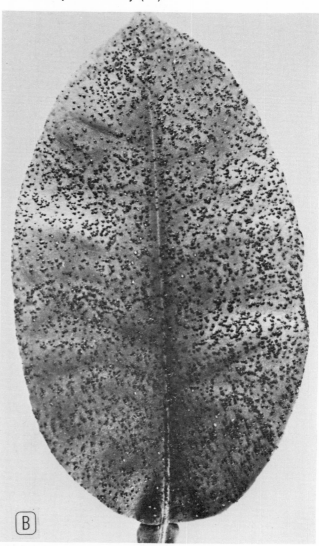

FIG. 39. *MELANOSE.* A. Translucent depressions on leaf surface—the first symptom of infection. B. Fully formed pustules.

Twig infections (Fig. 41A) are similar in appearance and development to those on leaves. Severe infections may cause young shoots to die back. Other parts of the tree that may be attacked are fruit pedicels and calyx lobes.

Cause

Melanose is caused by the fungus *Phomopsis citri* Fawc., known by its ascigerous stage as *Diaporthe citri* (Fawc.) Wolf. The Phomopsis stage, which pro-

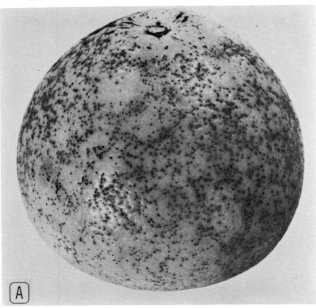

FIG. 40. *MELANOSE.* A. Separate pustules on grapefruit. B. Confluent pustules producing sharkskin and tearstain blemishes.

vides the chief source of inoculum, sporulates in dead twigs. Greatest sporulation occurs in tissues that died 2–4 months earlier, but some spore formation takes place in woody tissues that have been dead for as long as 2 years (281). The minute fruiting bodies are present throughout the year, but they do not discharge spores in midwinter or during extended periods of drought. Spores are extruded during rains, forming cream-colored masses at the tips of the black, flask-shaped fruiting bodies. Spores extruded after rains or during moist periods collect in soft stringy tendrils which on drying become hard and brittle. Spores are disseminated mainly by drops of rain or dew that wash over and drip from the pycnidia. Strong winds accompanying rains may blow spore-carrying drops as far as 30 feet from the source of inoculum.

Germ tubes penetrate young tissues by the secretion of pectic enzymes. No spores are produced by the melanose pustules while host tissues remain alive. This signifies that affected fruits on the tree are not a source of contagion. It is difficult to isolate the causal fungus from pustules or underlying host tissue, presumably because the fungus is killed by the gumming and enzymatic defenses of the host (6).

Spores capable of causing infection are also produced by the perfect stage of the fungus (280, 372). Ascocarps are formed in decaying wood in contact with the soil; these fruiting bodies forcibly eject ascospores into the air where they may be carried considerable distances by air currents. Because ascospores are produced in relatively small numbers, they are not a major cause of infection (281).

Environmental conditions affect both spore numbers and host susceptibility. The amount of inoculum is approximately proportional to the quantity of recently killed wood in the canopy. Killed wood in turn results from such factors as excessive shading, cold injury, storm damage, root rot, deep plowing, insect damage, and malnutrition. The susceptibility of host tissue is governed by its age: it is most susceptible shortly after leaves emerge and fruits set. By the time leaves are fully enlarged they are no longer susceptible. Fruits are liable to infection from the time of setting until they reach a certain size which varies with variety: oranges become immune after they reach a diameter of 1½ in., grapefruits after reaching 3 in. (281).

Other conditions contributing to epiphytotics are rainfall and temperature. Rains provide the vehicle for carrying spores from fruiting bodies to susceptible tissue and for providing conditions necessary for germination. Temperatures indirectly affect the emergence of new growth. With respect to the fungus, the temperature optimal for mycelial growth is 76°F, for pycnidial formation, 68–76°, and for conidial formation, 68° (377).

Young trees are less affected by melanose than older trees, presumably because of the smaller amount of killed wood in the canopies.

Control

Although infection may take place any time of the year when inoculum, susceptible host tissue, and rains coincide, it produces economic damage only during flushing and fruit setting in spring.

As far back as 1896 (323), it was shown that a single spray of Bordeaux mixture would control melanose if applied within several weeks of petal drop. But the particulate nature of the residue on the foliage and fruit led to serious increases in scale infestations (371). The difficulty was overcome when 1 per cent oil was mixed with the 6-6-100 Bordeaux mixture (369). Today, fixed or neutral copper sprays have supplanted Bordeaux because of ease in preparation, and scale control is being provided by improved scalicides. The recommended spray for melanose control is 3.75 lb. of copper (actual metallic content)/500 gal. water, applied 1–3 weeks after petal fall. In situations where melanose is serious year after year, or at times when springs are exceedingly wet, or when blooming is late or staggered, a second application is suggested 4 weeks after the first.

Because copper intensifies melanose pustules already formed (see **Star Melanose**), a search has long been underway for a more suitable fungicide. Of the possible substitutes tested to date, none has given as reliable results as copper (53, 316). Because copper is accumulating in toxic concentra-

FIG. 41. *MELANOSE.* A. Pustules on young shoots. B. Leaf distortion produced by a severe attack.

tions in the soils of more and more groves (see **Copper Toxicity**), the search goes on for non-copper-containing fungicides (229, 317, 319).

Though pycnidia are borne in dead wood, little control has resulted from pruning alone, mainly because the bulk of inoculum comes from small dead twigs and fruit spurs that escape the pruner. The removal of dead wood lessens the incidence of melanose only in years when trees are extensively damaged by freezes (281).

MESOPHYLL COLLAPSE. A leaf disorder first evident as angular translucent chloroses and later as parched areas within the blade.

Some years many leaves develop an angular chlorosis between lateral veins and from the midrib toward the margin (Fig. 42). The area appears water-soaked and translucent when observed with transmitted light, and the leaf surface over the chlorosis is slightly wrinkled. With time, affected areas turn gray to light brown and, when invaded by secondary fungi (e.g., *Alternaria*, *Cladosporium*, or *Colletotrichum*), become brownish-black and studded with fungus fruiting bodies.

The cause of mesophyll collapse is still debatable. At one time, red spider mite (*Paratetranychus citri* [McG.]) and sand blasting were thought to produce the trouble by erosion of the cuticle or waterseal of the leaf. Investigations have since shown that the cuticle remains intact and that the trouble arises from within the leaf (332). Analyses of the mesophyll sap from affected areas reveal a lower calcium content and a higher magnesium, phosphorus, potassium, and sodium content than sap from normal tissue. These findings suggest that mesophyll collapse may be due to a weakening of the normal gel structure in the mesophyll; this results from a decrease in calcium brought about by its removal through replacement with monovalent cations. It is thought that this imbalance is related to excesses or deficiencies in soil constituents. Contributing causes may be drying winds, inadequate irrigation, or poor root systems. At present there are no assured ways of preventing mesophyll collapse.

MOLYBDENUM DEFICIENCY. See Yellow Spot.

MOMPA. See Felt.

MUSHROOM ROOT ROT. A root decay and top decline caused by certain mushroom-producing fungi. *SYN.*—Clitocybe root rot, Armillaria root rot, oak root fungus disease.

Two mushroom-producing fungi are known to cause decline and death of citrus trees. One, *Armillaria mellea* (Vahl) Kummer, is destructive in

FIG. 42. *MESOPHYLL COLLAPSE.* Affected areas are at first water-soaked; later they dry out and become papery.

certain areas of California and Australia and has been reported on citrus also in Texas, Corsica, Italy, Malta, Cyprus, Tunisia, Malawi, and Rhodesia (189). The other, *Clitocybe tabescens* (Scop.) Bres., has been mentioned on citrus only in Florida and Morocco (52). However, both fungi are so similar in behavior and pathogenicity that advantage can be taken of the literature on Armillaria root rot when dealing with Florida's Clitocybe root rot problem.

Distribution

C. tabescens occurs from Florida west to Texas, Oklahoma, and Missouri, and north to Virginia. *A. mellea* is also widely distributed and in some areas it is coextensive with *C. tabescens*. In Florida, *A. mellea* has been reported on noncitrus plants from the northern part of the state.

Importance

C. tabescens is destructive to some 210 species of plants belonging to 137 genera in 59 families (271). In addition to citrus, it attacks plantings of peach, guava, lychee, loquat, banana, grape, and tung.

When reckoned in terms of total citrus acreage, the incidence of mushroom root rot is low, but within certain localities it may be high, some groves being veritable hotbeds of infection. Greatest incidence is found in the well-drained sandy soils of the Ridge. The disease rarely occurs in flatwoods, probably because oaks and other susceptible hardwoods are not naturally abundant in such areas prior to land clearing. Owing undoubtedly to the predominant use of rough lemon as a rootstock on sandy soils, the disease is seen most often in trees on this stock. In Florida there are few records of the disease on other stocks, though in Morocco, it is principally the sour orange that is attacked (52).

Aboveground lesions of mushroom root rot are frequently confused with those of foot rot; therefore, the importance of mushroom root rot has often been underestimated (365). The disease occurs mainly in Polk County but is occasionally encountered in the counties of Marion, Lake, Highlands, and Brevard. Trees decline and become unproductive.

Symptoms

The most conspicuous symptom of mushroom root rot is tree decline but since decline may also result from other agencies, this symptom in itself is of little diagnostic value. The distribution of declining trees, however, does afford some assistance in identification. Generally, stricken trees are scattered in restricted portions of a grove. In areas affected by mushroom root rot, trees decline and die intermittently over the years; therefore, such areas often contain large numbers of replants of various ages.

The only other aboveground symptom is an occasional rotting of the bark at the base of the trunk

(Fig. 43). Though similar in appearance to foot rot, trunk lesions of mushroom root rot differ in arising always from well belowground whereas foot rot lesions do not usually extend much below ground level. Mushroom root rot lesions generally extend up the trunk only a few inches from the ground. In old lesions, the bark is rotted away, affording entry to wood-decaying fungi such as *Ganoderma* spp. Symptoms of decline do not develop until the disease has killed a considerable portion of the root system.

It is belowground that reliable diagnostic symptoms are found. Some of the lateral roots, and usu-

FIG. 43. *MUSHROOM ROOT ROT*. Three stages of mushroom development of *Clitocybe tabescens* growing at the base of an infected citrus tree. Arrow points to a bark lesion originating from an infected root.

ally the taproot, are rotted for various distances. Occasionally, only one side of a root is involved and the lesion may be delimited and callused over.

Presence of the causal fungus provides further evidence that mushroom root rot is involved. When bark of a dead root is peeled back, the mycelium can be seen as thin, filmy wefts or leathery sheets, white when freshly developed or buff when old (Fig. 44). At the advancing margin, sheets are usually fan-shaped while the older parts are often perforated by closely set holes 0.5–1.0 mm in diameter. The sheets are coextensive with the dead bark except at the advancing margins. Freshly dug roots containing the living fungus have a pronounced mushroom odor.

Infection generally occurs along the lateral roots and progresses both away from and toward the root crown. In advanced cases, the fungus girdles the root crown and taproot, causing tops to decline quickly.

In some cases, superficial, blackish, hardened outgrowths (xylostroma) occur as extrusions through longitudinal fissures in the bark of attacked roots (Fig. 45). These outgrowths appear also in Armillaria root rot, and some investigators regard them, though incorrectly, as rhizomorphs (82, 93). The true shoestring-like rhizomorphs usually produced by *A. mellea* have not been seen under Florida conditions.

Further diagnostic evidence is provided by the characteristic fruiting structures of the causal fungus arising from trunk lesions and from affected roots near the trunk (Fig. 46). Given conditions favorable for their development, mushrooms appear in late summer and fall. Within several days, however, mushrooms wither to a blackish mass in which condition they may remain discernible for months.

Mushrooms of *C. tabescens* occur in clusters of several to many individuals arising from a common base. Caps are whitish to tan, convex at first, later flattened or centrally depressed, and 2–3 in. in diameter; gills are whitish and not covered by a membrane or veil and stems are without a ring or annulus.

The presence of either fresh or old clusters of these mushrooms makes possible the detection of mushroom root rot long before tops of trees decline or lesions develop on the trunks.

Cause

Controversy has long prevailed as to whether *Clitocybe tabescens* (Scop.) Bres. is the same as *Armillaria mellea* (Vahl) Kummer. Both fungi commonly occur saprophytically on stumps and roots of dead trees, and both act as parasites when roots of planted trees come in contact with such sources of infection. According to some taxonomists, each should be regarded as a distinct species since the mushrooms of *A. mellea* possess veils and annuli whereas those of *C. tabescens* possess neither. To other mycologists who regard these structures as variable and so of little taxonomic significance, *C. tabescens* is considered to be merely a nonannulate form of *A. mellea*. An inter-

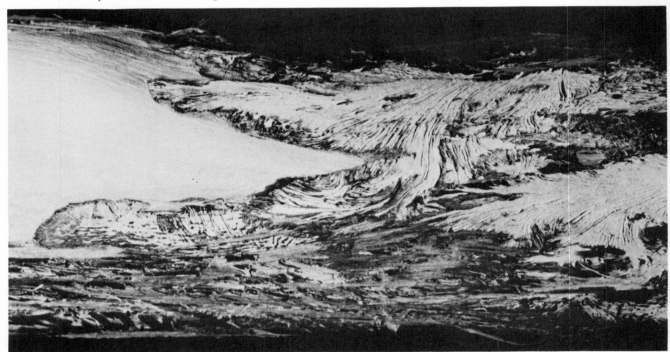

FIG. 44. *MUSHROOM ROOT ROT*. Whitish mycelial mats produced by *Clitocybe tabescens* beneath the rotting bark of a rough-lemon root. The fungus weft resembles a many-fingered glacier.

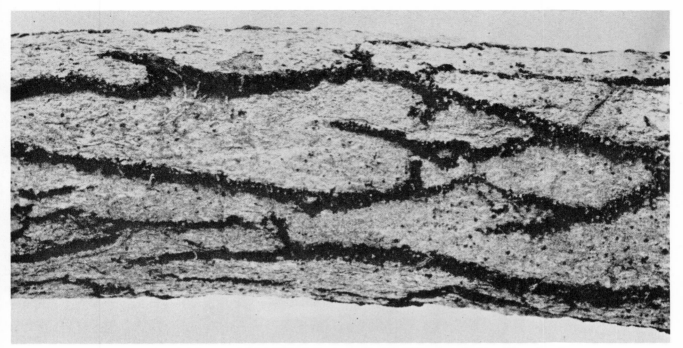

FIG. 45. *MUSHROOM ROOT ROT.* Root of sweet-orange seedling attacked by *Clitocybe tabescens,* showing black xylostroma extrusions through longitudinal cracks in the bark. Slightly enlarged. (Courtesy of A. S. Rhoads.)

FIG. 46. *MUSHROOM ROOT ROT.* Fruiting bodies of *Clitocybe tabescens* originating from an infected rough-lemon root. (Courtesy of J. O. Whiteside.)

mediate position is taken by those holding both fungi to be members of the same genus though differing at the species level. Viewed in this light, the two are classified either as *Armillariella tabescens* (Scop. ex Fr.) Sing. and *Armillariella mellea* (Vahl) Karst. or as *Armillaria tabescens* (Scop. ex Fr.) Emel. and *Armillaria mellea* (Vahl) Kummer (298).

In the absence of mushrooms, it is not possible in most cases to differentiate between Clitocybe and Armillaria root rots. In culture, there is much overlapping of the two species with respect to formation of rhizomorphs and temperature constants (131, 132). However, it has been pointed out (270) that many cultures of *A. mellea* exhibit luminescence whereas those of *C. tabescens* never do, and that the latter fruits readily in culture whereas most attempts to induce *A. mellea* to fruit have failed.

The sources of inoculum are stumps and roots of trees, mostly hardwoods and especially oaks, that remain in the ground after land has been cleared. As long as food reserves remain in root debris, the fungus survives, sometimes for many years, and is in a position to infect healthy citrus roots on contact.

Attempts to inoculate citrus roots, both wounded and noninjured, with pure cultures of *C. tabescens* have failed to reproduce the disease (271), but attempts have succeeded when naturally infected roots were placed in contact with healthy roots. Under grove conditions, injuries do not seem required for infection judging from the ease with which the fungus spreads from tree to tree by root contact.

Control

Outbreaks of mushroom root rot can be minimized at the time land is cleared for citrus by removing all stumps and roots. Once the disease attacks citrus roots, control is difficult.

Trenching and soil fumigation are used in California to combat the spread of Armillaria root rot into surrounding unaffected areas. For the fumigation of replant sites, a fumigant is injected into the soil to weaken the fungus and to render it susceptible to attacks by another soil-inhabiting fungus, *Trichoderma viride* (81). Soil fumigation for the control of mushroom root rot has not been tested under Florida conditions where modifications of the California procedures may be required owing to differences in soil types, moisture levels, and microflora.

MYCOIDEA. Genus of the Algae. One species, *M. parasitica* Cunningham (=*Cephaleuros virescens* Kunze), causes **Algal Disease** (which see).

MYCOSPHAERELLA. Genus of the ascomycetous fungi. *M. citri* Whiteside is the cause of **Greasy Spot** and **Greasy Spot Rind Blotch** in Florida.

MYIOPHAGUS. Genus of the phycomycetous fungi. *M. ucrainicus* (Wise) Sparrow is parasitic on scale insects. See **Entomogenous Fungi.**

MYRIANGIUM. Genus of the ascomycetous fungi. *M. duriaei* Mont. & Berk. and *M. floridanum* Hoehn. overgrow dead scale insects. See **Entomogenous Fungi.**

NAILHEAD RUST. See **Leprosis.**

NECTRIA. Genus of the ascomycetous fungi. *N. diploa* Berk. & Curt. and *N. coccophila* (Tul.) Wr. overgrow dead scale insects. See **Entomogenous Fungi.**

NEMATODES. A group of unsegmented, cylindrical worms; those associated with citrus are too small to be seen with the naked eye.

The two most important nematodes attacking citrus in Florida are *Radopholus similis* and *Tylenchulus semipenetrans* which are described, respectively, under **Spreading Decline** and **Slow Decline.** The following discussion concerns other species known to parasitize, or to be associated with, citrus in Florida.

Proof of pathogenicity to citrus has been obtained for *Belonolaimus gracilis* Steiner (303), *Trichodorus christiei* Allen (302), *Pratylenchus brachyurus* (Godfrey)—the cause of citrus slump— (32, 115, 327), and *Hemicycliophora arenaria* Raski (336). No information is yet available, however, regarding the economic effects of these species.

Other nematodes have been found associated with Florida citrus; some of them are *Xiphinema americanum* Cobb, *X. chambersi* Thorne, *X. coxi* Tarjan, *X. vulgare* Tarjan, *Pratylenchus penetrans* (Cobb), and *P. zeae* Graham. There are reasons for believing that these species are also pathogenic to citrus, but their economic importance remains to be evaluated.

NEMATOSPORA. Genus of the ascomycetous fungi. *N. coryli* Peglion and *N. gossypii* Ashby & Nowell have been reported to cause inspissosis, or dry rot, of fruits that have been punctured by plant bugs (357). Infrequent.

OAK ROOT FUNGUS DISEASE. See **Mushroom Root Rot.**

OIL-GLAND DARKENING. A rind blemish caused by prolonged storage at low temperatures.

Under abnormally low storage and transit temperatures, grapefruit and tangelos undergo a darkening of the oil glands, resulting in a polka-dotted appearance of part or all of the rind. In severe cases, the tissues between the oil glands also become discolored and may collapse, imparting a scalded appearance. Severely affected fruit may develop a soft, watery, internal breakdown. For the avoidance of this disorder, fruits should be stored and shipped at recommended temperatures (301). See also **Chilling Injury.**

OLEOCELLOSIS. A rind breakdown following the rupture of oil glands. *SYN.*—rind-oil spot, green spot.

Peel oil liberated by insect or thorn punctures, hail, cold damage, or rough handling of turgid fruits causes oleocellosis, a superficial browning and sinking of rind tissue between the oil glands, causing them to stand out prominently (see title page). Affected areas may be up to ½ in. in diameter (Fig. 47). In limes and lemons, the blemish is often so conspicuous that fruits must be culled before shipment to the fresh market. Areas of green fruit blemished by oleocellosis do not degreen

properly and give rise to a condition on the rind known as green spot.

Oil glands of lemons can be punctured by pressures ranging down to 1 lb. As fruits on the tree dry off during the day, the amount of pressure required to release oil rises to a maximum of 16 lb. Economically objectionable amounts of oleocellosis develop with rind oil rupture pressures up

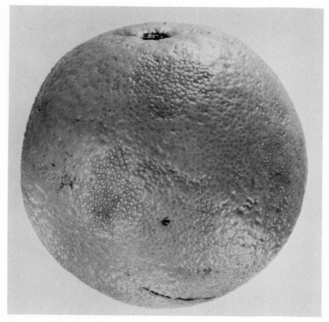

FIG. 47. *OLEOCELLOSIS.* Oil glands brought into relief by collapse of the rind. (Photograph by A. A. McCornack.)

to 7 lb. (247). Consequently, fruit should not be harvested while turgid or wet.

OOSPORA. See **Sour Rot.**

ORANGE LEAF SCAB. See **Scab.**

PELLICULARIA. Genus of the basidiomycetous fungi. *P. filamentosa* (Pat.) Rogers is one of the causes of **Damping Off** (which see).

PENICILLIUM. Genus of the ascomycetous fungi. *P. italicum* Wehmer and *P. digitatum* Sacc. cause respectively **Blue Mold** and **Green Mold** (which see).

PERCHLORATE TOXICITY. An orange-colored chlorosis of leaf tips caused by an impurity often present in natural sources of potassium nitrate. *SYN.*—yellow tipping.

The presence of perchlorate in fertilizer may lead to a chlorosis at the tips of older leaves (307). In mild cases, symptoms appear only on leaf surfaces exposed to the sun. In more severe cases, symptoms are present on both surfaces, may involve half the leaf area, and may be followed by defoliation. In contrast to the similar-appearing **Biuret Toxicity** (which see), perchlorate chlorosis

FIG. 48. *PERCHLORATE TOXICITY.* A. Tip chlorosis produced by perchlorate impurity when present to the extent of 1 per cent. B. Reduction of chlorosis resulting from improved refining methods that decrease perchlorate content to 0.3–0.4 per cent. (Photograph A by I. Stewart, B by R. L. Reese.)

is orange-colored and is usually traversed by green veins (Fig. 48A).

Perchlorate impurities occur in natural sources of the double salt of sodium potassium nitrate but not in synthetic potassium nitrate. Improved methods of refining have reduced perchlorate levels in Chilian nitrate of soda potash (263) to an extent that yellow tipping is hardly noticeable (Fig. 48B).

PESTALOZZIA. Genus of the deuteromycetous

fungi. *P. guepini* Desm. has been isolated from old melanose lesions (5, 6).

PHOMOPSIS. Genus of the deuteromycetous fungi. *P. citri* Fawc. (the imperfect stage of *Diaporthe citri* [Fawc.] Wolf) is the cause of **Melanose** (which see). See footnote under **Diaporthe.**

PHOMOPSIS STEM-END ROT. One of several fruit rots that originate at the stem end. *SYN.*—stem-end rot (a name also used for similar rots caused by Diplodia and Alternaria fungi).

The fungus that causes melanose in the grove causes a stem-end rot of fruits in the packinghouse. Fruits may decay at any time during the 21 days normally elapsing from harvest to consumption (35).

Symptoms

The rot starts around the button, producing at first a slight softening of the rind. The affected area increases rapidly and the fruit develops a pliable, tan- or brown-colored rot (Fig. 49).

Inside the fruit, the rot advances quickly through the core and along the inner wall of the peel. The flesh is also invaded to some extent, causing it to taste bitter.

It is difficult to distinguish the stem-end rot caused by the Phomopsis fungus from a similar rot produced by Diplodia fungi. Differentiation requires identification of the causal fungus. In late stages, Diplodia stem-end rot may be recognized by the development of decay at both ends of the fruit. The disease predominates in fruits that have been degreened with ethylene (31).

Cause

Phomopsis stem-end rot is caused by *Phomopsis citri* Fawc., a fungus known in its ascigerous state as *Diaporthe citri* (Fawc.) Wolf.

Spores of the fungus originate in recently killed wood (See **Melanose**). During rainy weather, spores are extruded from pycnidia and are washed over the fruit. Some become lodged under the calyx lobes where they are protected from desiccation. Histological examinations of the rind during various stages of fruit maturation have shown the spores and mycelial fragments to persist in a viable condition and to remain capable of initiating decay after fruits have been picked (156, 227). Occasionally, fruits may be attacked while still on the tree but usually only when they are injured or overripe. After harvest, the fungus enters the fruit through the area of abscission. Once within the rind tissue, the fungus grows rapidly in cells rich with starch and then progresses into the albedo. Symptoms of infection can be seen on the rind 2–4 days after the fungus has gained entry (35).

Control

In the packinghouse, development of the disease is checked by standard fungicidal treatments employing thiabendazole, diphenyl, or sodium

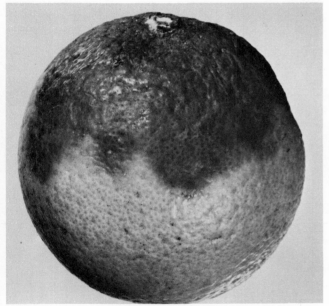

FIG. 49. *PHOMOPSIS STEM-END ROT.* A brown, pliable, postharvest rot. (Photograph by A. A. McCornack.)

orthophenolphenate (157). Benomyl promises superior control (221) but cannot be recommended until it has received official clearance. Postharvest fungicide treatments do not always provide satisfactory control, particularly when fruits are not treated until several days after harvest.

It has been demonstrated that a substantial reduction in the amount of stem-end rot developing in the packinghouse can be obtained by spraying trees with benomyl before harvest (33, 34).

PHYLLACTIDIUM. Genus of the Algae. One species, *P. tropicum* Mobius (thought to be synonymous with *Cephaleuros virescens* Kunze) causes **Algal Disease** (which see).

PHYLLOSTICTA. Genus of the deuteromycetous fungi. *P. hesperidearum* Penz. has been found in the necrotic tissue of leaf spots (93). Rare.

PHYSALOSPORA. Genus of the ascomycetous fungi. *P. rhodina* (Berk. & Curt.) Cke. (the ascigerous state of *Diplodia natalensis* P. Evans) may be found in weakened or injured tissues throughout the tree. In the packinghouse, it causes **Diplodia Stem-End Rot** (which see). *P. fusca* N. E. Stevens has been reported on twigs (5).

PHYTOPHTHORA. Genus of the phycomycetous fungi. *P. parasitica* Dastur (=*P. nicotianae* B. de Haan var. *parasitica* [Dastur] Waterh.) causes **Foot Rot** and **Brown Rot** (which see). *P. citrophthora* (Sm. & Sm.) Leonian also causes brown rot. The foregoing are the two species that attack citrus in Florida. Elsewhere in the world, citrus may also be attacked by *P. hibernalis* Carne, *P. boehmeriae* Sawada, *P. megasperma* Dreschsler, *P. palmivora* Butler, *P. syringae* Kleb., *P. citricola* Sawada, and *P. citri* Ven.

PHYTOPHTHORA GUMMOSIS. See **Foot Rot.**

PINHEAD YELLOW SPOT. See **Measles.**

PINHOLE ROT. See **Blue Mold, Green Mold.**

PINK FUNGUS. *Nectria diploa* Berk. & Curt., one of the so-called "friendly fungi." See **Entomogenous Fungi.**

PINK PITTING. A pink to dark-brown necrosis and pitting on areas of the rind between the oil glands. *SYN.*—pitting of grapefruit.

Grapefruit grown for the fresh market may fail to qualify as bright fruit because of a blemish called pink pitting (Fig. 50). The disease is of recent appearance in Florida and seems to be related to the replacement of oil sprays by some of the newer organic pesticides.

Pink pitting causes no problems when affected fruits are utilized for processing.

Pink pitting is faintly visible on fruits on the tree as early as September when portions of the rind show a pale, dusty discoloration that may be confused with feeding injuries of the rust mite. When viewed under the hand lens, rust mite damage appears as flat, brown dabs of varnish over the rind whereas pink pitting appears as shallow, pink depressions generally restricted to areas of the rind between the oil glands (see title page). Later in the development of pink pitting, some of the depressed areas enlarge to 0.5 mm in diameter and turn dark brown. The blemish is readily apparent in the field about the first of October.

All varieties of grapefruit are susceptible, and groves in all areas of the state may show the disease. Occurrence is sporadic, and incidence varies from year to year. In some groves, up to 90 per cent of the crop has been found affected (322).

A Cercospora-like fungus has been isolated from affected areas of the rind (322), but pathogenicity studies remain to be carried out before this isolate can be implicated.[1] The fungus has also been found in unaffected tissue, suggesting that environmental factors play a role in the development of pink pitting.

Control of pink pitting is reported from the use of various sprays including copper, zineb, dithianon, benomyl, Polyram, sulfenimide, and oil, but only copper (0.75 lb. metallic/100 gal. water), zineb (2 lb./100 gal. water), and oil (1.3 per cent), which have official clearance for use on citrus, can be recommended at this time (322). Copper or zineb may be combined with oil, but zineb–oil sprays are preferred because they avoid the development of **Star Melanose** (which see).

PITTING OF GRAPEFRUIT. See **Pink Pitting.**

1. J. O. Whiteside (Plant Disease Reptr. 56:671–75. 1972) reported that pink pitting is caused by *Mycosphaerella citri*, recently described as the fungus causing greasy spot (364). He considers that pink pitting is merely a more severe form of the rind stippling described under **Greasy Spot** (which see). The name **Greasy Spot Rind Blotch** has been proposed to cover all known symptoms on the rind caused by *M. citri*, including those described here under **Pink Pitting.** Whiteside also reported that satisfactory control of greasy spot rind blotch is obtained by the same fungicidal treatments that are effective against foliar greasy spot.

PLANT CITY DISEASE. See **Blight.**

PLASTER DISEASE. See **Felt.**

PODAGRA. A bud-transmissible disease of kumquat trees on rough-lemon rootstocks.

Podagra is found in Florida wherever certain clones of Nagami and Meiwa kumquats are budded on rough lemon (181). The disease does not appear when the same clones are budded on Cleopatra mandarin, sour orange, or sweet orange.

Podagra has destroyed up to 50 per cent of trees in some plantings. Notwithstanding such losses,

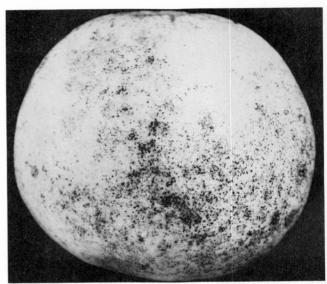

FIG. 50. *PINK PITTING.* A pink to dark-brown rind necrosis and pitting of grapefruit.

kumquat growers prefer using rough lemon as a rootstock because it produces lower solids, thus minimizing fruit splitting.

The name "podagra," from the Greek *pous* (foot) plus *agra* (seizure) refers to the swollen rootstock portion of the trunk. Overgrowth is accompanied by a scaling of the rough-lemon bark (Fig. 51). The overgrowth and scaling, together with a stunting and decline of the top, give the impression that trees are budded on exocortis-affected trifoliate orange. In spring, profuse gumming may accompany the scaling. Symptoms appear 1–10 years after budding. Rate of deterioration varies: some trees succumb rapidly in the nursery, others decline gradually in the field.

The possibility that podagra is a virus disease is supported by results of transmission tests showing that budwood taken from affected trees (but not from healthy ones) reproduces podagra when grafted into rough lemon (195).

PODONECTRIA. Genus of the ascomycetous fungi. *P. coccicola* (Ell. & Ev.) Petch overgrows dead scale insects. See **Entomogenous Fungi.**

PORIA. Genus of the basidiomycetous fungi. *P. cocos* Wolf has been reported to occur on decaying roots (5).

PRATYLENCHUS. Genus of phytophagous nematodes. *P. brachyurus* (Godfrey) (one of the lesion nematodes and the cause of citrus slump) has been proven to be pathogenic to citrus, but its economic effects remain to be evaluated. See **Nematodes.**

PSOROSIS. A group of virus diseases having in common a characteristic vein-banding chlorosis in immature leaves. *SYN.*—California scaly bark (but not Florida scaly bark, which is a synonym for leprosis). The name psorosis was originally coined for a disease known today as Rio Grande gummosis, but the name was later misappropriated and is currently used for the virus disease under discussion.

In 1954, 317 grove trees, selected more or less at random, were indexed under the Florida Budwood Registration Program. Of this number, 12 per cent were infected with the virus of psorosis (236). Today, the incidence of psorosis is undoubtedly lower as the result of propagations since 1956 with psorosis-free budwood.

The name psorosis applies to a group of diseases exhibiting different symptoms but having in common a type of vein-banding chlorosis in immature leaves. Diseases encompassed by this group are psorosis A, psorosis B, concave gum, and blind pocket. Some authorities also include **Infectious Variegation** and **Crinkly Leaf** (which see). The economic effects of infection with psorosis A are bark scaling and subsequent tree decline. Though capable of damaging trees, psorosis B, concave gum, and blind pocket are seldom encountered in Florida.

Plants affected

All varieties of citrus can carry the viruses that cause these diseases. Bark scaling and tree decline, however, have been reported only in trees of sweet orange, grapefruit, and mandarin.

With respect to plants that show leaf symptoms, most varieties develop the vein-banding pattern, but some express symptoms more readily than others (274).

FIG. 51. *PODAGRA.* Meiwa kumquat on rough lemon, showing swelling and scaling of rootstock portion of the trunk.

Symptoms

All types of psorosis are capable of developing vein-banding symptoms in leaves that are one-quarter to one-half mature. Two patterns are recognized: fleck and oakleaf. The fleck pattern consists of faint, water-white, ¼–1 mm wide bands on both sides of the veins and veinlets (Fig. 52A). The oakleaf pattern consists of a generalized chlorosis, usually in the shape of a leaf, superimposed over the central portion of the fleck pattern (Fig. 52B). In leaves of the spring flush, oakleaf patterns are conspicuous but in summer and fall flushes of the same tree flecking patterns replace the oakleaf patterns. Both fleck and oakleaf patterns are more or less symmetric with respect to the midrib, i.e., the pattern on one side of the blade is an approximate mirror image of that on the other side. In this regard, psorosis vein clearing differs from tristeza vein clearing; in the latter case, cleared areas are shorter and scattered at random. Fleck and oakleaf patterns are best seen by transmitted light. Symptoms occur most commonly in the spring flush and persist for a few days to several weeks. The symptoms are often present in only a few leaves on an infected tree.

Similar leaf patterns may result from injuries due to mites, aphids, blowing sand, and genetic abnormalities (342). These pseudo-psorosis symptoms can be recognized by their lack of bilateral symmetry and their nontransmissibility by budding.

Various strains of psorosis induce symptoms in mature leaves, trunks, and fruits. For convenience, these symptoms are described under the following types of psorosis.

Psorosis A.—This is the common form of psorosis. Bark scaling develops from 6 to an indefinite number of years after the appearance of leaf symptoms. Cases are known of infected trees remaining productive for 40 years before onset of scaling and decline. Scaling begins in isolated areas of the trunk or main limbs and consists of the separation from the bark of small, dry, irregular flakes 1/12–1/8 in. thick. A cut across the bark at a tangent to the flaked area shows underlying tissues to be discolored. Scaling is progressive and involves more and more of the bark until the affected area encircles the trunk (Fig. 53B) and limbs (Fig. 53A). Unless rubbed off, scales remain attached at the periphery of the lesion and curl outward in a pagoda-roof fashion. A small amount of gum may appear around the edges of the shelled area. Underneath the scaling, young xylem tissue is impregnated with brown gum. Some of the subsequently deposited layers of xylem in turn be-

FIG. 52. *PSOROSIS*. Young-leaf symptoms. A. Fleck pattern. B. Oakleaf pattern. C. Unaffected leaf for comparison.

FIG. 53. *PSOROSIS.* A. Limb scaling. B. Trunk scaling. C. Tree decline.

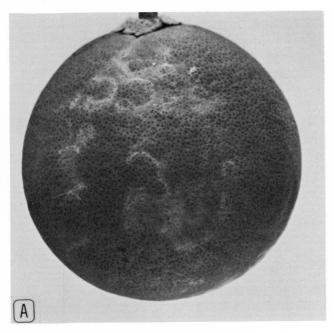

FIG. 54. *PSOROSIS.* A. Symptoms of psorosis B on the rind of grapefruit. B. Mature-leaf symptoms of psorosis B on grapefruit.

come impregnated, so that eventually the wood—as seen in cross section—contains crescent-shaped bands of brown wood alternating with bands of unaffected wood (355). Once the xylem is affected, the flow of water is impeded and, as a consequence, leaves wilt and drop and the top declines (Fig. 53C).

Psorosis B.—Scaling and wood symptoms are similar to psorosis A except that they develop more extensively. In further contrast to psorosis A, psorosis B produces symptoms on fruits and mature leaves. Fruits may show partial or complete rings of sunken tissue on the rind (Fig. 54A), and leaves may show chlorotic blotches that contain greasy-spot-like swellings on the undersurfaces (Fig. 54B) (183).

Blind pocket.—One strain of psorosis causes narrow, lens-shaped depressions paralleling the long axis of the trunk (Fig. 55A). The bark over the cavities remains intact. Under the cavities, woody tissues are impregnated with a waxy material, and a series of gum layers extend into the wood from the base of the pocket.

Concave gum.—Another strain of psorosis causes pockets that are longer, wider, and shallower than those of blind pocket (Fig. 55B). Concave gum may lead to considerable trunk and limb distortion. Some affected trees are stunted. The growth habits of certain varieties, e.g., mandarins, can produce trunk irregularities that may be confused with blind pocket.

Symptoms other than those given for the above-mentioned diseases may appear during greenhouse testing to determine the presence of psorosis virus in suspect trees. The first pinpoint flush of leaves that develops after grafting may drop, and the first shoots may die back. Succeeding flushes and leaves are normal except for fleck and oakleaf patterns.

Cause

The various forms of psorosis result from infection by different strains of the psorosis virus group. Transmission of these strains is by budding. No vectors are known. Mechanical transmission has not yet been reported (342, 343),[1] but limited dodder

1. Unpublished research has been mentioned (343) as indicating sap transmissibility of the virus of either psorosis A or blind pocket.

transmission is known to occur (260, 353). Seed transmission takes place to some extent in certain clones of Carrizo citrange (30, 62) and Troyer citrange (262).

The strain producing oakleaf patterns is associated with concave gum psorosis. The strains producing fleck are associated with psorosis A, psorosis B, or blind pocket.

If tissue for grafting is taken from within a scaling lesion, the receptor seedling rapidly develops the bark shelling described for psorosis B. If, on the other hand, graft tissue is taken outside the scaling lesion of the same tree, the receptor seedling does not develop scaling symptoms until several years later. Inoculum from both lesioned and non-lesioned bark produces young-leaf flecking symptoms within 6 months of grafting. These reactions suggest that the viruses of psorosis A and psorosis B are merely components of the psorosis A strain and that, therefore, the term psorosis B is redundant (341).

In indexing procedures, the use of sweet orange, mandarin, and lemon seedlings is recommended. Symptoms in immature leaves appear 1–6 months following grafting. Leaf patterns do not develop properly under high temperatures.

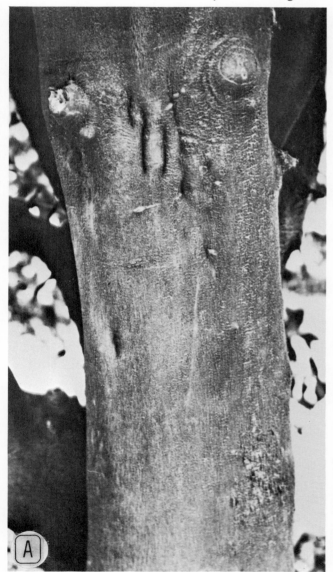

FIG. 55. *PSOROSIS.* A. Blind-pocket psorosis. B. Concave-gum psorosis. (Photograph A by J. M. Wallace.)

For the purpose of providing psorosis-free bud-wood, the presence of young-leaf symptoms on a bud source tree or the appearance of leaf patterns in indicator plants is sufficient evidence to disqualify that source tree. For the identification of strains involved, however, much longer periods of testing are required (342, 343).

Control

Once a tree becomes infected with psorosis virus, it cannot be cured. The productive life of a scaling tree may be prolonged by painting early stage lesioned areas with DN-75 (169), once a widespread practice in California. It is now considered preferable to pull affected trees once the cost of maintenance exceeds returns, and to replant with virus-free trees.

PUFFING. A disorder of fruits on the tree in which the rind is irregularly thickened.

Puffing is often used interchangeably with **Creasing** (which see). In puffing, however, the irregularity of the rind results from an abnormal thickening of certain parts of the albedo whereas in creasing the irregularity is due to a collapse of separated cells in the albedo. Commercially, the external effects are much the same, but physiologically, puffing and creasing, though sometimes found on the same fruit, appear to be unrelated (164). The causes of puffing and their control are not understood.

PYTHIUM. Genus of the phycomycetous fungi. *P.* sp. has been isolated from plants affected by **Damping Off** (which see).

QUICK DECLINE. See **Tristeza.**

QUIESCENT INFECTIONS. See **Symptomless Infections.**

RADOPHOLUS. Genus of nematodes. *R. similis* (Cobb) Thorne, 1949, is the cause of **Spreading Decline** (which see).

RAGGRINZIMENTO DELLA BUCCIA. See **Rumple.**

RED ASCHERSONIA FUNGUS. *Aschersonia aleyrodis* Webber is one of the so-called "friendly fungi." See **Entomogenous Fungi.**

RED FUNGUS. *Sphaerostilbe aurantiicola* (B. & B.) Petch (syn. *S. coccophila* Tul.) is one of the so-called "friendly fungi." See **Entomogenous Fungi.**

RED RUST. A name applied to two different diseases: **Exanthema** and **Algal Disease** (which see).

RHIZOCTONIA. Genus of the deuteromycetous fungi. *R. solani* Kühn (the imperfect stage of *Pellicularia filamentosa* [Pat.] Rogers) is one of the causal organisms of **Damping Off** (which see).

RICING. See **Granulation.**

RIND BREAKDOWN. See **Chilling Injury.**

RIND-OIL SPOT. See **Oleocellosis.**

RIND PITTING. See **Chilling Injury.**

RIO GRANDE GUMMOSIS. One of several gumming disorders of the bark. *SYN.*—gummosis, Florida gummosis, Diplodia gummosis, tears, ferment gum disease, gum disease.

Gumming of the bark, like watering of the eyes, is a normal reaction to various excitants—mechanical, physiological, and pathological. The bark of citrus gums whether injured by implements, deranged by malnutrition, or infected by certain fungi, bacteria, or viruses. Most types of gumming can be attributed to known causes, but Rio Grande gummosis remains unaccounted for.

Rio Grande gummosis is a recent name for an old disease (133). In 1896 (323), the trouble was designated psorosis, a term that has subsequently been misappropriated and is now being used for a disease of virus origin. Another old name, gummosis, has so long been used indiscriminately for any gumming of the trunk that today it has lost its specific meaning. Hence, the more recent name Rio Grande gummosis is adopted to bring back into focus the particular gumming disease under discussion.

Plants affected

Rio Grande gummosis is seen most frequently in trees of lemon and grapefruit, occasionally of orange, and seldom of tangerine. Trees on rough lemon rootstocks appear to gum more profusely than those on sour orange.

Importance

As recently as 1931, the disease was regarded as one of the more important citrus problems in Florida (272). Today, it is still encountered but can no longer be considered prevalent.

Symptoms

Rio Grande gummosis usually affects trees after they have come to bearing age. In lemon trees, gumming breaks out just above the bud union and involves progressively more and more of the trunk, reacting in these respects like shell bark. In grapefruit and orange trees, symptoms appear higher up the trunk and out on the larger branches.

First symptoms consist of a slight vertical cracking of the bark (usually in the center of a water-soaked area) and a release of pale yellow gum that runs down the trunk. The gum may be washed away by rain or it may harden and accumulate over the bark in large masses (Fig. 56A). The exudation takes place principally in spring and early summer. Removal of bark from the gummed area discloses small, green, woody galls projecting from the woody cylinder (54). Upon the healing of a young lesion, a thin layer of dead bark is sloughed off, exposing a buff-colored scar (Fig. 56B). Healing, however, may be only temporary; later, scars may again exude gum, reheal, and in the process enlarge the scaled portion of the trunk. With repeated scaling, areas become quite large and expose the wood. Gum pockets found in the wood are shallow in young lesions, deep in old. In old infections, the wood shows a buff discoloration in which are imbedded concentric orange rings (Fig. 57). The gum pockets cause the outer layers of the wood as well as the bark to bulge. On release of the gum, the pockets become cavities (in size up to 0.5 x 2.0 in.) studded with gall tissue.

The symptoms of Rio Grande gummosis (RGG) resemble in several respects those of foot rot (FR) and psorosis (P), but the three diseases can be dif-

FIG. 56. *RIO GRANDE GUMMOSIS.* A. Initial gumming symptoms. B. Scars left by healed lesions. (Photographs by J. F. L. Childs.)

ferentiated on the basis of the following reactions. *Gumming*: RGG, very copious; FR, copious; P, inconspicuous. *Portion of bark sloughed*: RGG, outer layers; FR, entire bark; P, outer layers. *Gum pockets in wood*: RGG, present; FR, absent; P, absent. *Color of affected wood*: RGG, salmon in a banded pattern; FR, yellow to brown; P, brown (53).

Cause

Pathogenic organisms have long been suspected as the incitants of Rio Grande gummosis. Inoculations with *Diplodia natalensis* (91) and *Phomopsis citri* (304, 305) have led to gumming but not to the reproduction of typical gummosis lesions. It remains possible that these secondary invaders play a role in enlarging the original lesions. In California, ferment gum disease (synonymous with Rio Grande gummosis) has been attributed to infection by a basidiomycetous fungus (38).

There is some circumstantial evidence that the disease is related to high concentrations of chlorides in the soil (63).

Control

Preventive measures are not known. Surgical treatments serve no useful purpose and may even weaken the tree. Exposed wood can be painted with a wound dressing like DeKaGo to prevent invasion of wood rot fungi. The remaining alternative is to replace affected trees when they become nonproductive.

ROADSIDE DECLINE. See Blight.

ROBINSON DIEBACK. A twig dieback frequently encountered in Robinson tangerine trees. *SYN.—* twig dieback of Robinson tangerine, Diplodia twig dieback.

Trees of Robinson tangerine often develop a gumming, drooping, wilting, and dieback of twigs on scattered branches. At times, the disease progresses downward from the tips to involve the branches. The dieback may even proceed into the trunks of young trees and cause the death of entire plants. Up to 6 in. of growth may die within a week's time. The maximum amount of dieback is observed in the fall.

Isolations from affected twigs yield *Diplodia natalensis* along with other fungi usually present in decaying citrus tissue, and reinoculation of Robinson plants in the greenhouse with *Diplodia* reproduces the condition (98). It is reported that the spraying of Robinson trees in the field during March and June with benomyl (5 oz. of active in-

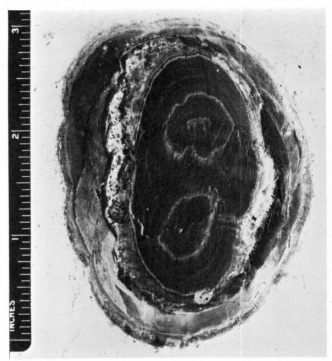

FIG. 57. *RIO GRANDE GUMMOSIS.* Tangential section across the wood of an old trunk lesion. Orange-colored rings alternate with buff-colored ones. (Photograph by J. F. L. Childs.)

gredient/100 gal. water) leads to considerable reduction in the incidence of dieback (153).

Diplodia natalensis is known to be present on citrus trees without causing disease (98, 277), and the fungus is also known to colonize weakened tissue. Pruning off the affected limb well below the diseased portion does not necessarily stop further dieback of that limb (98). Observations suggest that the Robinson top has a higher water requirement than can be met at all times by the rootstock.

RUMPLE. A preharvest disorder of lemon fruits characterized by a network-like settling of the rind between the oil glands; first apparent at time of colorbreak. *SYN.—*wrinkle-rind, raggrinzimento della buccia, mangiato d'agro.

Rumple is a rind collapse of lemons that prejudices the sale of fruits on the fresh market. Blemished fruits, however, can be processed without affecting the taste of lemonade concentrate and the quality and yield of peel oil.

Plants affected

Rumple has been found in all types of true lemons grown in Florida. It appeared in a varietal trial at Avon Park in each of 38 selections under test, including Villafranca, Eureka, Lisbon, Bearss (Sicilian), Avon, Harvey, and Arizona. Not known to be affected are rough lemon and such hybrids as Meyer and Perrine (205). In Italy, rumple occurs in the lemon varieties Femminello Oliva and Do.-Sa.Co. (284, 285).

Trees of grapefruit, sweet orange, and tangerine growing next to rumple-bearing lemon trees show no abnormalities; neither do fruits of these 3 varieties when borne on sprouts arising from interstocks of trees topworked to rumple-producing clones of lemons (205).

History and distribution

The malady was first encountered in 1956 in the aforementioned varietal trial (187). It has subsequently been seen in most of the state's lemon plantings.

Abroad, rumple occurs in Sicily (284), Cyprus (205), and Ethiopia (205), and, judging from descriptions and photographs, also in Israel, Lebanon, and Turkey. It was not encountered, in the course of limited observations, in Spain, Morocco, Egypt, South Africa, and India (205).

Importance

In Florida, up to three-quarters of the crop in certain plantings has been found affected. In Italy, packouts have been reduced by as much as 30 per cent (284) and in Turkey, 75 per cent (172).

Losses may be minimized, however, by a judicious handling of the crop. In lemons grown for the fresh-fruit market, rumple may be avoided by picking fruits before rind collapse commences, i.e., before the onset of colorbreak. When picked after the development of rumple, no loss results if affected lemons are processed since such fruits do not alter the taste of lemonade concentrate or the quality and quantity of recovered peel oil

(172). Rumple is important only when it occurs in tree-ripened lemons intended for the fresh-fruit market.

In storage, affected areas of the rind may increase in size and new lesions may develop in previously sound fruits, especially if fruits are held at high temperatures and low humidities; below 68°F, however, postharvest involvement is of little commercial importance. Decay is not conspicuously greater in rumpled fruits than in sound fruits (248).

Symptoms

Rumple first appears in late summer when fruits turn yellow. The earliest symptom is a chlorotic speckling of the rind. When viewed under a hand lens, each speck consists of a yellow, slightly sunken area involving 4 or 5 oil glands, the yellow discoloration being restricted to flavedo tissue between the glands. With increase in size and depression of an affected area, walls of glands undergo a series of color changes, becoming greenish-brown, then tan, mahogany-brown, and finally, brownish-black (Fig. 58C). Glands ultimately flatten and collapse.

Lesions progress from specks to large irregular areas showing a net-like pattern suggestive of worms tunneling beneath the rind (Fig. 58A). After fruits break color, some affected areas remain yellow-brown whereas others turn green to greenish-black.

With time, more and more of the rind becomes involved, but further extension of lesions practically ceases after fruits reach maturity.

In certain years, affected areas degenerate further to produce black necrotic spots within the primary net-like pattern—a condition referred to as "secondary rumple" (Fig. 58B). Such lesions are hard and rarely admit secondary soft-rot organisms even after 30 days in storage. Areas of secondary rumple range up to 1 in. in diameter.

Rumple causes no appreciable fruit drop, and it does not affect foliage or the growth of trees.

Cause

Studies in Florida and Sicily have as yet failed to identify the causal agent. That rumple might be the result of a somatic mutation or bud sport is suggested by (a) the often-noted restriction of affected fruits to single limbs and (b) the finding that the percentage of affected fruits in a given tree remains approximately the same year after year (205). It is unlikely, however, that the same mutation would occur simultaneously in all 38 selections

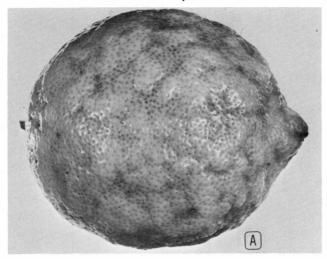

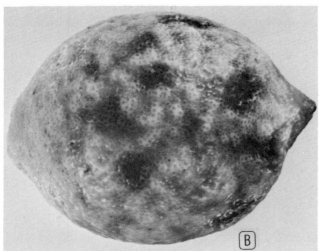

FIG. 58. *RUMPLE.* A. Primary stage. B. Secondary stage showing development of black, hard spots within areas of primary rumple. C. Tangential section of lemon rind showing discolored walls of oil glands under area of primary rumple.

of the above-mentioned varietal trial or in each of the nucellar clones of Ross, Galligan, Ledig, Price, Prior, and Monroe lemons that have been imported into Florida from California.

Repeated isolations from diseased tissues have failed to implicate a fungus or a bacterium (205). No control of rumple was obtained by spraying trees with neutral copper (53 per cent, 1.4 lb./100 gal. water) at 6-week intervals from March to July.

Rumple was not prevented by spraying trees with azinphosmethyl, Parathion, ethion, or chlorobenzilate, thus suggesting that the inciting agent is neither a mite nor an insect (205).

That the cause might be a virus has been investigated by trials in Florida (205) and Sicily (285), but results to date are inconclusive. From observational evidence, however, the virus hypothesis appears untenable. For example, the Bearss lemon in Florida develops considerable rumple in some plantings but not in others, despite the fact that all budsticks have been derived from a common source. Furthermore, buds of nucellar Lisbon types imported from California, where rumple is not known to occur, have in Florida produced trees bearing affected fruits.

In Florida, no consistent differences in the content of N, P, K, Ca, Mg, Mn, Zn, Fe, Cu, and B were found in rumple-affected and normal fruits along with subtending leaves borne on adjacent limbs of the same tree, though in Sicily, high K was found to correlate with low rumple (286). In Florida, soil applications at very high rates of N, P, K, Ca, and Mg failed to affect significantly the amount of rumple, though there have been some indications that high rates of K tend to decrease the percentage of affected fruits (205, 286) and that high rates of N increase the percentage (205). Although fertilizer treatments at times modify the percentage of affected fruits, they do not overcome completely the seemingly genetic basis of rumple.

Rumple has not responded to supplemental irrigation or to applications of gibberellin and antidessicant waxes (205).

Present indications suggest that rumple results from a genetic sensitivity in certain clones of lemons to some as yet undetermined environmental factors.

Control

Without knowing the cause of rumple, no control recommendations can be made at this time except that rumple-free clones should be selected for propagation. In affected groves, losses may be minimized by picking lemons for the fresh-fruit market before colorbreak, i.e., before the commencement of rumple. Fruits harvested after colorbreak can be utilized by the processor for the manufacture of frozen lemonade concentrate and for the recovery of oil without prejudice to the finished products.

RUSSET. A blemishing of the rind by the citrus rust mite *(Phyllocoptruta oleivora)*, resulting in a lowering of external and internal fruit quality. *SYN.*— rust mite injury.

Despite its entomological origin, russet is included here because it can be confused with melanose. The sense of touch usually distinguishes the two: fruits with russet have a smooth feel whereas those with melanose have a sandpapery texture. When both blemishes occur on the same fruit, a hand lens is needed to identify the two: the damage due to rust mite is flat (varnish-like), diffuse, finely grained, and brown, whereas the damage due to melanose consists of reddish-brown or black pustules surrounded by borders of buff-colored scar-like tissue (see title page). Diagnosis becomes more difficult when fruits have been affected by rust mite and melanose at an early stage in their development. Early rust mite injury produces a buff- or olive-colored coating ("shark-skin") that with expansion of fruit develops slight cracks (141). Melanose of the mudcake type also produces a coating but it is reddish to dark brown in color, cracked into rather large platelets, and often studded with blackish pustules. The recognition of these two blemishes is at times complicated by the presence of other types of rind speckling such as **Pink Pitting, Greasy Spot,** and **Flyspeck** (which see).

Russet is controlled by miticides whereas melanose, pink pitting, and flyspeck are controlled by fungicides.

RUST MITE INJURY. See Russet.

SALT BURN. A scorching of plant tissues by high concentrations of salt in the soil or in seaspray and overhead irrigation water.

Protoplasm is coagulated and killed by excess salts whether taken up from the soil or deposited by seaspray or by brackish water from overhead irrigation systems. When uptake through the roots is chronic, foliage is dull green, bronzed, and thin; when root uptake is acute, leaves develop a sudden tip or marginal chlorosis followed by necrosis (Fig. 59) and shedding. In severe cases, defoliated twigs die back and fruits are dropped. When salts are deposited externally, the symptoms are similar to those of acute uptake by the roots.

High concentrations of salt in the soil may result from natural saline conditions, from intrusions of sea water into waters used for irrigation, or from excess applications of fertilizer. Reduction of toxic concentrations in the soil may be accomplished by flushing with fresh water.

Water testing between 1,000 and 1,500 ppm total dissolved solids may produce burn when applied through overhead sprinkler systems at low rates and during daytime hours when evaporation is rapid. The same water, however, may be used safely for high-rate, short-duration sprinkling or for flood irrigation (48). Water testing above 1,500 ppm should not be applied through overhead systems.

SANDHILL DECLINE. A disease of mature trees (especially those growing in sandy soils) that is characterized by foliar chlorosis, dieback, and fruit dwarfing. *SYN.*—greening-like disease, blight, and young-tree decline.

Since 1967, many trees in the 10–30-year age group and mainly on rough-lemon root have declined in the Ridge section, with incidence greatest in Hardee and Highlands counties (194). Affected trees show most of the symptoms commonly associated with **Young-Tree Decline** (which see). Leaves appear deficient in zinc and are speckled with green dots (Fig. 60A); leaves subsequently formed are small, leathery, erect, and chlorotic (Fig. 60B); trees die back; and about 2 per cent of the crop consists of mature but golf-ball-sized, distorted fruits containing aborted seed and curved columellas (Fig. 61). The disease, tentatively named sandhill decline, differs from young-tree decline in appearing later in the life of the tree and in not being preceded by a chronic wilting of the foliage.

Some investigators consider sandhill decline to be identical with, or a variation of, **Blight**[1] (which see). The two diseases might best be considered distinct until their etiologies are proved identical. No measures are known that will restore the health of diseased trees. Trees no longer productive should be removed, but it is not yet certain whether replants will grow normally.

SCAB. An infectious disease producing corky eruptions on fruits, leaves, and stems. SYN.—sour-orange scab, orange-leaf scab, lemon scab, grapefruit scab, verrucosis.

There are three different fungi that cause scab on citrus. Only one, namely sour-orange scab (caused by *Elsinoë fawcetti* Bitanc. & Jenkins), occurs in Florida. The second, called sweet-orange scab (caused by *Elsinoë australis* Bitanc. & Jenkins), is restricted to South America; if introduced into Florida, it would cause serious losses to the fresh-fruit grower (186, 189).[2] The third, known as Tryon's scab (caused by *Sphaceloma fawcetti* var. *scabiosa* [McAlp. & Tryon] Jenkins), is presently confined to Australia, New Zealand, New Caledonia, New Guinea, Sri Lanka, Rhodesia, South Africa, and Argentina (162).

Sour-orange scab is a problem in the growing of certain citrus fruits for the fresh market and in the rearing of certain rootstock varieties in the nursery.

1. In his 1936 publication on blight, Rhoads (269) described a disease that may well constitute the first account of sandhill decline: "A distinctly different form of decline of citrus trees on rough lemon stock, which has become quite widespread and prevalent in parts of the ridge section of the state during the last few years, is erroneously termed 'blight' by many growers who are not familiar with the latter trouble. Unlike typical blighted trees, these declining trees do not exhibit a chronic wilting of the foliage, but, instead, a very pronounced mottle-leaf or frenched and small-leaved condition, especially on the ends of the branches."

2. During the past decade, sweet-orange scab was intercepted at U.S. ports of entry over 1,000 times.

Plants affected

Varieties of fruits that are very susceptible to scab include Temple, King, Murcott, lemon, sour orange, Clementine mandarin, and some of the tangelos (e.g., Minneola, Thornton, Sampson, and Webber). Less susceptible (though at times sufficiently so to cause economic loss) are grapefruit, Satsuma mandarin, tangerine, Tahiti lime, and various of the tangelos (e.g., Orlando). Rarely affected are sweet-orange[3] and Nagami kumquat. Varieties considered immune are Key lime, limequat, Marumi kumquat, and citron (370). Within a variety, there may be degrees of susceptibility; thus, in grapefruit, Royal and Triumph (possibly hybrid types) and Marsh Seedless are less affected than Duncan, Hall, Conner, and Foster.

In the nursery, scab attacks seedlings of sour orange, rough lemon, Rangpur lime, trifoliate orange, and certain of the citranges (e.g., Rusk and Carrizo). The tender apical growth is affected (Fig. 63), resulting in stunted, bushy plants.

Symptoms

On fruit.—When fully formed, scab lesions are raised and range in color from buff through pink to olive-drab (depending on stage of development and on secondary fungi found overgrowing lesions). Eruptions occur either as individual round pimples (Fig. 62A) or, on the confluence of many infection points, as sheet-like patches (Fig. 62B). In many varieties (e.g., tangelos, lemons, and sour oranges) scabs are perched on conical projections of the rind, while in other varieties (notably grapefruit), scabs coalesce and flatten into extensive sheets. With enlargement of the fruit, the sheets crack into small polygonal platelets (Fig. 62B) and impart a texture to the rind similar to that produced by wind damage, thrips injury, and mudcake melanose.

Fruits heavily infected may drop shortly after being attacked whereas those remaining on the tree

3. Of late, puzzling outbreaks of scab have appeared on sweet oranges in the Indian River and Immokalee areas of Florida. The corky eruptions resemble those caused by sour-orange scab and are clearly not those produced by sweet-orange scab. The possibility is being investigated that a new strain of *E. fawcetti* may be involved.

may be scarred and distorted to such a degree that they become unmarketable as fresh fruit. Scab does not invade the flesh, and lesions do not provide an entryway for secondary fungi causing fruit rot. Scab does, however, decrease the recovery of peel oil in such varieties as lemon and Tahiti lime.

Scab may sometimes be confused with similar effects produced by other agencies. It can be differentiated from canker by the cauliflower-like contour of the scab surface (see title page). Scab resembles melanose in being raised but differs from it in forming light-colored and much larger protuberances than the brownish-black, pinhead-sized pustules of melanose.

On leaves.—First evidence of infection consists of minute, circular protuberances usually on the undersides of young leaves. A few days later these projections develop a cream to yellow-orange color at the tips. With expansion of the leaf, lesions become more conspicuous, forming conical outgrowths with corresponding depressions on the other surface of the leaf. Severely attacked leaves become distorted (Fig. 63) and occasionally drop prematurely.

FIG. 59. *SALT BURN.* A. Necrosis due to excess sodium absorption. B. Necrosis due to excess chloride absorption. (Photographs by H. D. Chapman.)

On shoots and young stems.—The watersprouts of trees and the stems of nursery plants develop scab lesions similar to those on the leaves. Other parts of the tree on which scab lesions may be found are tender twigs, blossom pedicels, and buttons. The economic effect of scab in the nursery is to stunt seedlings and to make them difficult to bud because of their bushiness.

Cause

In the older literature, scab was erroneously attributed to *Cladosporium citri* Massee. Not until 1907 was the true pathogen discovered (89). It was later named *Sphaceloma fawcetti* Jenkins on the basis of the imperfect stage (160) and *Elsinoë fawcetti* Bitanc. & Jenkins when the perfect stage was discovered (22). Studies in Trinidad suggest that there are different strains of the scab organism and that the one attacking grapefruit may be a mutant of the one causing scab on other varieties (12).

The incidence of scab is a function of interacting factors which include degree of varietal susceptibility, presence of host tissue in a juvenile state, inoculum potential, water for spore dispersal and germination, and temperature. The role played by varietal susceptibility has been discussed above.

Tissues are most susceptible at the time leaves, fruits, and shoots emerge. Resistance increases with age. Fruits are no longer susceptible after reaching a diameter of ¾ in. (367).

The number of spores available to penetrate susceptible tissue determines the number of lesions produced. Conidia arise from the surfaces of scabs formed the preceding season (161, 376). They are disseminated by dripping water and to some extent by wind and insects. Being minute in size,

FIG. 60. *SANDHILL DECLINE.* A. Green-dotted areas in frenched leaves. Similar dots may be caused by insect punctures and melanose pustules. B. Dwarfed, leathery, strap-shaped, green-speckled, chlorotic leaves.

spores may be suspended for long periods in atmospheric vapor (376). The fungus lives from one season to the next on infected leaves and, to a lesser extent, on infected fruits. The role that spores of the perfect stage play in the development of the disease is not understood.

Water is the most important single factor affecting severity of the disease (376), being involved in both spore dissemination and germination. Water, however, need not be in the form of rain or irrigation—scab is often severe during periods of minimum rainfall. Heavy dews or fogs are believed to function equally well in spore dispersal and germination. Scab occurs annually in groves on damp, low-lying hammocks and flatwoods; it occurs only to a negligible extent in groves on the Ridge (282).

Temperature optima of the scab fungus are: for conidial formation, 68–80°F; for germination, 56–90°; and for shortest incubation (approximately 5 days in very young leaves), 56–86° (376). Thus, the

FIG. 61. *SANDHILL DECLINE.* Top: Mature, but golf-ball-sized, distorted fruit containing aborted seed and showing curved columella. Bottom: Normal-sized fruit from the same tree.

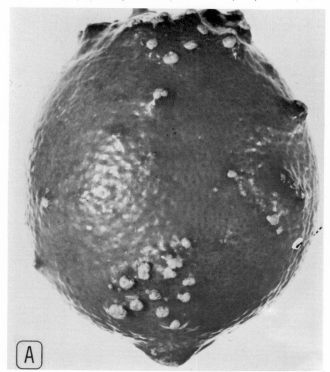

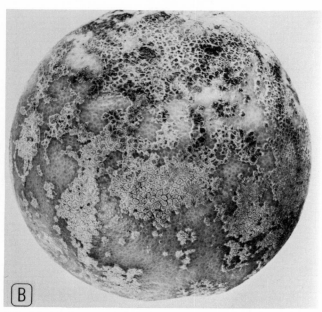

FIG. 62. *SCAB.* A. Perched lesions on lemon. B. Confluence of multiple infections to produce sheets of scurfy rind on grapefruit.

scab fungus can infect fruits in summer as well as in spring. The economic importance of summer infection is negligible, but the biological importance may be considerable—fruits set in summer are usually left on the tree and, if infected, may carry the fungus over winter and provide inoculum the following spring.

Control

Present recommendations for the control of scab are the application of at least two fungicidal sprays, the first just in advance of the spring flush and the second when two-thirds of the petals have fallen. Either ferbam (7.5 lb. 75 per cent wettable powder or 6.0 lb. 95 per cent wettable powder/500 gal. water) or fixed copper (3.75 lb. metallic content/500 gal. water) may be used. Ferbam has been found more effective than copper for the two-thirds petal-fall application. In situations where scab is a perennial problem, the strength of copper

FIG. 63. *SCAB*. Lesions causing distortion of leaves. Death of the apical portion of the stem stimulates side branching and results in the formation of a bushy plant difficult to bud.

for the prebloom application should be increased 1.5–2.0 times. Where extra-clean fruit is desired, an additional spray should be applied 2–3 weeks after blossoming (104, 109).

Unusually effective control of scab has been reported from a single application of benomyl. The systemic action of this fungicide appears to overcome the difficulty of spraying large acreages within the short sporulation-infection period of the fungus (151, 152).

A reduction in spore load and carry-over of the fungus can be expected from the removal of rough-lemon and sour-orange rootsprouts underneath trees (282).

The control of scab in the nursery is given under **Alternaria Leafspot** (which see).

SCALD. See Chilling Injury.

SCALY BUTT. See Exocortis.

SCLEROTIUM. Genus of the deuteromycetous fungi. *S. rolfsii* Sacc. has been reported to occur on fruit touching the ground (93, 301) and to be one of the causes of **DAMPING OFF** (272). Rare.

SEPTOBASIDIUM. Genus of the basidiomycetous fungi. *S. pseudopedicellatum* Burt and *S. lepidosaphes* Couch cause epiphytic growths on bark, leaves, and fruits of citrus in the Gulf states. See **Felt.**

SHEAF NEMATODES. See Hemicycliophora, Nematodes.

SHELL BARK. A scaling of the outer bark on trunks of lemon trees; usually accompanied by intermittent gumming induced by secondary fungi. *SYN.*—decorticosis.

Trunks of most lemon trees in Florida develop a scaling of the outer bark after trees become 7 years old or more. The appearance of shell bark can be alarming, but effects on cropping and longevity are difficult to assess since few nonaffected trees exist on which to base a comparison. Despite profuse scaling, 60-year-old lemon trees have been known to remain commercially viable in

FIG. 64. *SHELL BARK*. A. Scaling of the outer bark above the bud union in a 12-year-old Villafranca lemon tree. B. Scaling in the lemon portion of a topworked grapefruit tree. The bud unions are at the base of the shelling.

Florida. The disorder occurs in many parts of the world, but for some inexplicable reason it is unknown in certain areas (173).

Plants affected

In general, clones of Eureka lemon are quickest to show symptoms and those of Lisbon (except the open type and USDA semidense strains) much slower, though differences among selections of a particular variety are recognized. In Florida, the widely grown Bearss lemon begins to scale within 10 years. The disorder has rarely been seen in hybrid lemons such as the Meyer.

Symptoms

Shell bark causes a scaling of the outer bark, first evident usually just above the bud union. In time, the entire trunk and the larger limbs may be covered with scales. Trees that have been topworked to lemon show scaling in the bark of lemon branches but not in the bark of the sandwiched varieties (Fig. 64B). Clones with greatest susceptibility begin to scale at about 7 years from budding; those tolerant may not scale until trees are more than 25 years old. The outer bark dries out, shrinks, and cracks into platelets and vertical strips (Fig. 64A). The inner bark and cambium normally remain intact, and the affected outer bark is sloughed after healing occurs. In humid areas, shelling provides infection courts for a number of weakly pathogenic fungi (e.g., *Alternaria citri, Colletotrichum gloeosporioides, Diaporthe citri, Diplodia* spp., and *Fusarium* spp.) that stimulate gumming and prevent healing (43). There appears to be a cyclic susceptibility of tissue: once regenerated bark becomes 5–7 years old, it may again start shelling.

Histological examinations of the tissues under scaled areas have shown that the disease begins and spreads in the layers of the middle cortex and occasionally extends into conductive tissues (36). At first, the exterior bark appears sound, but ultimately it dries out and dies in areas overlying the inner disorganized, brownish tissue.

In the absence of trees without shell bark to serve for comparison, the attribution of top symptoms such as chlorotic foliage, leaf drop, and bare twigs to shell bark is not convincing.

Cause

Shell bark has been attributed variously to a fungus *(Diaporthe citri* [Fawc.] Wolf) (92), to a non-infectious degeneration (36), and in part to a virus, possibly exocortis (45). The causes of the disease are probably complex and are not yet fully understood.

Control

No curative or preventive measures have been worked out in Florida. In California, the hatracking of affected trees has been reported to hasten the healing of trunk lesions and to effect temporary recovery. Preventive measures found promising include: (a) the selection of budwood for propagation from old trees that are free of shell bark; (b) the practice of budding high, which delays and minimizes the inception of shell bark (25); and (c) the propagation of tolerant selections like the various Lisbon clones (particularly the Keen, Monroe, Price, and Prior) that do not develop shell bark until trees are rather old (20).

SLOUGHING. A postharvest rind breakdown of grapefruit.

Sloughing disease occurs sporadically in very early pickings of red and pink grapefruit. Its occurrence has been limited to the Ridge and West Coast districts. Though at times present in all fruits of a single picking, sloughing may fail to develop in fruits of succeeding harvests from the same tree.

First symptoms consist of a paling of areas on the rind (often in a tearstain pattern) followed by a rapid, chocolate-brown discoloration, and death of rind and albedo tissues. In initial stages, sloughing does not affect the firmness of the fruit or its flavor. About 5 days following first symptoms (usually about the time fruit reaches the market), the discolored areas become soft and moist and with handling slip off the sound tissues beneath (Fig. 65). The flesh continues to remain sound. Sloughed fruits are seldom overrun by green and blue molds.

The cause of sloughing is unknown. Apparently it is not produced by pathogenic organisms. Suspected as causes are physiological immaturity and picking during periods of high temperatures and prolonged rains (140). Sloughing has almost disappeared in commercial shipments of Florida grapefruit since the raising of maturity standards in 1967. The development of the disorder is checked by prompt precooling of fruit and subsequent refrigeration in transit.

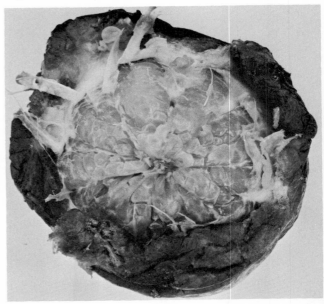

FIG. 65. *SLOUGHING.* Grapefruit in advanced stage, showing slippage of rind away from intact flesh. (Photograph by W. Grierson.)

SLOW DECLINE. A gradual deterioration of trees caused by the citrus-root nematode. SYN.—citrus nematode disease.

The two major nematode diseases of citrus in Florida are spreading decline and slow decline. The effects of the former are dramatic and obvious, those of the latter, insidious and often difficult to discern. Some authorities suspect that the widespread slow decline is exacting a greater toll in Florida citrus production than the more restricted spreading decline.

Of 1,580 groves examined in 26 Florida counties up to 1967, 53 per cent were found to be infested with the citrus nematode (325). As of 1962, infestations were recorded in 12 per cent of 2,500 citrus nurseries (146).

Plants affected

The citrus-root nematode, unlike the polyphagous burrowing nematode, appears to affect mainly *Citrus* and closely related genera. Within this group, however, there are wide differences in susceptibility (8, 97). Among the most resistant rootstock varieties are trifoliate orange and certain of the citranges (49). Sweet lime and rough lemon are more susceptible than sour orange (77, 241).

Varietal resistance is reported to be associated with a hypersensitive reaction to the feeding of the nematode, to the formation of wound periderm in the root cortex which walls off lesioned tissue, and to a toxic factor in the juice of roots (337).

Other than rutaceous hosts, the citrus-root nematode has been found to parasitize grape, lilac, persimmon, olive, cabbage palm, climbing hempweed (*Mikania batatifolia*), and a grass (*Andropogon rhizomatus*).

Symptoms

Although symptoms of decline appear in the tops of trees, these are not specific for the nematode. Symptoms are similar to those produced by any agent that injures roots. The foliage is dull green to bronze, leaves are erect and cupped, new growth lacks vigor, and trees wilt before adjacent healthy ones in times of drought. In late stages of decline, leaves are noticeably smaller and tops are much thinner than those of normal trees. Eventually, size and yield of fruit are reduced to uneconomic levels. As trees become older, symptoms become more pronounced. In extreme cases, trees are rendered worthless but are not killed. Affected trees occur in groups but, unlike spreading decline, there is no perceptible advancing margin. Replants do not grow off well in infested sites (145).

An examination of roots is required for the diagnosis of slow decline. Parasitized feeder roots are encrusted with sand that cannot be shaken off. The coating gives rootlets the appearance of being stouter than normal (Fig. 66), but no swellings or galls are produced. Because other factors may cause sand particles to adhere, this symptom alone is not dependable. Feeder roots that are severely attacked tend to be more stubby, misshapen, and brittle, and they lack the usual yellow or white color of normal roots.

Microscopic examination of rootlets is required to provide definite evidence of infection. Under 10X magnification, nests of female nematodes (up

to 100 per 4 mm section of rootlet) and clumps of eggs can be seen protruding from the surface of the roots. The head of the female is embedded near the pericycle of the cortex, and the swollen posterior portion remains outside the rootlet (Fig. 67A). Early larval stages (Fig. 67B) and males are 280–340 microns long by 10–17 microns wide. Larvae that will become females may or may not be

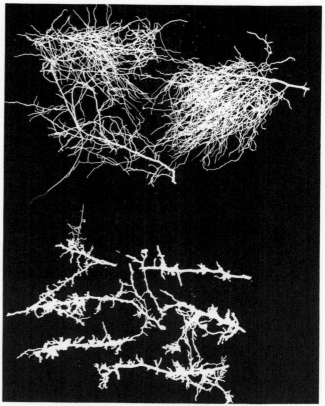

FIG. 66. *SLOW DECLINE.* Top: Normal noninfected feeder roots. Bottom: Feeder roots infected by the citrus nematode. (Photograph by C. I. Hannon.)

found attached to roots. Males, on the other hand, do not penetrate citrus roots.

Histopathological symptoms consist of a thickening of the cytoplasm in attacked cells of the cortex, an enlargement of nuclei and nucleoli, a decrease in vacuolar space, and the presence of wound cork around unsuccessful penetration sites (292).

Effects of the nematode on fruit production are difficult to measure. Considerable time is usually required for the increase of high nematode concentrations. Peak populations may not occur until 12–17 years after trees have been planted in infested soil (78). Trees do not decline markedly until the quantity of nematode larvae and males are in excess of 40,000 per 10-gram root sample (77). Under conditions similar to those in Arizona, trees may not show symptoms in the tops until 3–5 years after roots have become damaged (268). In advanced stages of tree decline, nematode populations decrease for lack of an adequate food supply in the form of new feeder roots.

Cause

Slow decline results from the parasitism of citrus by *Tylenchulus semipenetrans* Cobb. Two biotypes have been described from California on the basis of differential reactions on citrus (11). Another biotype isolated in Florida from *Andropogon rhizomatus* does not attack citrus (310).

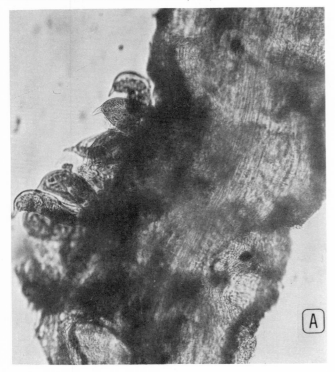

FIG. 67. *SLOW DECLINE.* A. Females of *Tylenchulus semipenetrans* protruding from side of feeder root. B. Juvenile stage of the citrus nematode. (Photograph A by A. C. Tarjan; B after Cobb [67].)

Some evidence points to the complicity of certain soil fungi (e.g., *Fusarium solani*) in colonizing feeding sites and furthering root destruction (75, 339).

Only females or larvae that will become females feed on citrus; juvenile females may be free living while the larvae are in search of a feeding site. The greatest number of penetrations occur in rootlets 3–4 weeks old; few penetrations occur after rootlets reach the age of 9 weeks (74). The minimum time required for completion of the life cycle varies with the host variety: at 75°F, it takes 14 weeks for completion on trifoliate orange, and 7 weeks on sour orange (76). Soil temperatures affect the activity and development of nematodes (243). At 75°F, eggs hatch in about 14 days (335). The optimum temperature range for infection is 77–88°F. Only slight penetration occurs at 59°F and at 95°F. Below 59°, the nematode becomes quiescent but does not die. In moist soil under laboratory conditions, nematodes remain alive 2.5 months at 91°F, 6.5 months at 81°, less than 1 year at 70°, and more than 1 year at 48–59° (7). Under Florida conditions, soil temperatures optimum for the development of citrus nematodes occur the year round.

In California, the citrus nematode was found to persist for 9 years after removal of infected citrus trees and after subsequent cropping with nonhost plants (10). In Florida, the nematode has been found to survive 18 months in moist soil in pots and 31 months in field plots free from citrus feeder roots (147). Even in water, nematodes may survive for 180 days (230). Frequency of irrigation and rainfall influence the length of the life cycle by flushing juveniles from egg masses into the soil from where they migrate in search of new feeding sites (335). Soil structure also influences the rate of reproduction: the rate was found to be significantly lower in soils containing 50 per cent clay than in soils with 5, 15, and 30 per cent clay. The highest rates of nematode reproduction and plant retardation were found to take place in soils containing 10–15 per cent clay (338). Organic amendments to the soil favor an early increase in nematode populations and thus cause early damage to citrus plants (241).

Once a nematode larva becomes imbedded in

the rootlet, it matures to adulthood and remains there until it dies. New penetrations of roots are made by juvenile nematodes. Juveniles move only very short distances and at a rate one-sixtieth to one-eightieth as fast as those of the burrowing nematode (145).

Spread of the nematode is principally through movement of infected nursery stock; they may also be disseminated to some extent by flowing surface water, cultivation equipment, and movement of fill soil.

Control

The way to prevent slow decline is to plant nematode-free nursery stock in nematode-free soil or in soil fumigated before planting. Once the nematode becomes established in a grove, control is difficult and costly.

For fumigating land preparatory to planting, use is recommended of a mixture of dichloropropane and dichloropropene (DD) at the rate of 60 gal./acre injected 12–16 in. deep with chisels set at 18 in. intervals. After injection, the furrows should be sealed with a cultipacker or roller. The waiting period before planting is reckoned at 1 day/gal. of the material used. Alternatively, 1-3 dichloropropene (Telone) may be applied in the same manner but at a rate of 48 gal./acre.

Infested replant sites should be treated with nematicides before planting with nematode-free nursery stock. Effective for the purpose is 98 per cent methyl bromide (available in 1 lb. containers) released into the soil through an applicator device. Injecting to a depth of 1 foot gives effective control of nematodes in the top 4 in. of soil within a 4-foot radius and at the 18-in. level within a 5-foot radius. Preliminary data show a 28 per cent increase in trunk circumference 1 year after application of this treatment (242).

It is possible to increase the yield of infected trees by fumigating the soil around trees in place. The use of 1,2-dibromo-3-dichloropropane (DBCP, Nemagon, or Fumazone) at 52–69 lb. active ingredient/acre, applied with soil chisels or through perforated irrigation pipe, has been found to reduce nematode populations and to increase fruit sizes and yields from 10 to 26 per cent over a 2-year period. Adequate soil moisture, elimination of weeds, and proper preparation of the soil are prerequisites for effective control (244, 326). Not all nematodes are killed by this treatment and it is necessary to re-treat from time to time. Proper intervals remain to be determined (244).

On the basis of long observation, some Florida authorities believe that infected trees can be restored to a reasonable degree of health through supplemental irrigation and nutrition—practices that stimulate the formation of new rootlets to replace those weakened by the citrus nematode.

Effective disinfection of nursery stock without phytotoxic effects has been reported from a 30-minute immersion in a 250–600 ppm solution of ethyl 4-(methylthio)-m-tolyl isopropylphosphoramidate (245).

Because of the rapidity with which new fumigants and applicators are being developed and because of frequent changes in regulations governing the use of pesticides, growers should consult specialists for latest recommendations.

SLUMP. A decline of citrus trees caused by *Pratylenchus brachyurus* (Godfrey). See **Nematodes.**

SOOTY BLOTCH. A cloudy blemish of the rind produced by the growth of a superficial fungus.

Sooty blotch is a rind blemish of minor importance caused by *Gloedes pomigena* (Schw.) Colby. It differs from **Sooty Mold** (which see) in forming dark, cloudy, circular colonies up to ½ in. in diameter and in not requiring insect-secreted honeydew for its development. The superficial colonies are composed of dark strands of the nonparasitic fungus and are at times intermixed with the fungus causing **Flyspeck.** Though the sooty-blotch fungus is susceptible to copper sprays, its control in the field is seldom warranted. If occasionally lots of fruit with sooty blotch reach the packinghouse, they can be brightened by the addition to the wash water of a mixture of calcium hypochlorite and sodium bicarbonate (334).

SOOTY MOLD. A superficial, easily detachable, black film of fungus growth that develops on fruit and leaves following the deposition of insect-secreted honeydew.

Sooty mold is more a signal than a disease. Its development on fruit and leaves (Fig. 68) indicates there has been laxity in the control of aphids, mealybugs, whitefly, brown soft scale, black scale, and other honeydew-secreting insects. The fungi that utilize honeydew as a source of food do not infect citrus though the fungal coating may interfere somewhat with the photosynthesis and gas ex-

FIG. 68. *SOOTY MOLD.* Black fungal growth on citrus fruit. Its superficiality is apparent from the flaking of the sooty covering.

change of leaves. Heavy deposits have been suspected of reducing flowering and fruit production. On fruit, sooty mold has been reported to retard maturation and to cause uneven coloration. Removal of the black film constitutes an expense in preparing saleable produce for the fresh fruit market.

Many fungi are known that can grow in honeydew (148). In Florida citrus, the species most commonly encountered are *Capnodium citri* Berk. & Desm. and *C. citricola* McAlp.

The amount of sooty mold on trees is roughly proportional to the population of honeydew-secreting insects present. These insects, and in turn sooty mold, can be controlled insecticidally or biologically.

If removal of the black film on trees is desired, spray oil can be applied to loosen the mold and allow it to be flaked off by wind and rain.

SOUR-ORANGE SCAB. See Scab.

SOUR ROT. A mushy, sour-smelling decay of fruit in storage and transit. *SYN.*—oospora rot.

Fruits long in storage are subject to a soft, readily smeared decay due to *Geotrichum candidum* Link (formerly *Oospora citri-aurantii* [Ferr.] Sacc. & Syd.). Varieties most often affected are Dancy mandarin, Temple orange, lemon, and other specialty fruits. First symptoms are water-soaked, dark yellow, slightly raised areas on the rind which soon become covered by a thin, creamy layer of fungus growth. The breakdown first affects the rind and segment walls and later also the juice vesicles. With further decomposition, fruits turn watery and start to leak and emit a sour odor that attracts flies and leads to the presence of maggots. The rot spreads by contact from fruits initially inoculated through rind injuries or through weakened areas around the buttons.

Losses from sour rot can be reduced by the prevention of injuries to the rind, by the elimination of overmature fruit, and by the avoidance of overlong storage (300). Control is not provided by thiabendazole (TBZ).

SPACKLE. See Tar Spot.

SPANISH MOSS. A bromeliaceous, nonparasitic epiphyte infesting citrus trees and producing pendulous masses of stringy, grayish-green vegetation. *SYN.*—Florida moss.

Spanish moss, *Tillandsia* (=*Dendropogon*) *usneoides* L., is neither from Spain nor a moss. It is a member of the pineapple family and indigenous to the American tropics. Though a misnomer, it does spell neglect. Since this air plant is intolerant of copper, its presence in citrus trees indicates that annual copper sprays have not been applied for the control of melanose, scab, or greasy spot. Once leaf-dropping diseases thin the canopy, Spanish moss proliferates on branches exposed to sunlight (Fig. 69).

Spanish moss is spread by wind-borne and bird-carried vegetative fragments and by seed. At first the plant is affixed by weak roots but these soon dry up; from then on attachment is provided by the twisting of the basal part of the strand around the twig of the tree.

Except for its affront to the senses, its excessive shading, and its resistance to high winds, Spanish moss is harmless. It is not a parasite and uses citrus merely as a platform on which to lead an independent existence. Besides citrus, Spanish moss grows on many other trees (but not palms) and on inanimate surfaces such as fenceposts and insulated (but not bare) electric wires.

Other bromeliaceous relatives of Spanish moss may also be found on citrus trees. These likewise

FIG. 69. *SPANISH MOSS. Tillandsia usneoides* L. is an epiphyte on citrus.

are nonparasitic. Members of the Bromeliaceae thrive under humid conditions and are found most frequently in hammock groves.

Infestations can be removed by a moss puller or by spraying with either fixed copper (3.75 lb. metallic/500 gal. water) or lead arsenate (2 lb./100 gal. water). Since copper-containing sprays discolor painted surfaces, it is preferable to use lead arsenate in the vicinity of buildings, and since lead arsenate is toxic to cattle, to use copper sulfate near pastures. Florida law prohibits the sale of arsenated fruit from all but grapefruit trees. Sprays must wet the clumps thoroughly. It may take several years for wind to blow out the dead moss.

Reinfestations can be prevented by cultural practices that maintain citrus trees in full canopy.

SPECKLE. See **Measles.**

SPHACELOMA. Genus of the deuteromycetous fungi. *S. fawcetti* Jenkins (the imperfect stage of *Elsinoë fawcetti* Bitanc. & Jenkins) is the cause in Florida of **Scab** (which see).

SPHAEROPSIS. Genus of the deuteromycetous fungi. *S. tumefaciens* Hedges (=*Haplosporella* of some authorities) is the cause of **Sphaeropsis Knot** (which see).

SPHAEROPSIS KNOT. A fungus-induced galling of limbs resulting in tree decline; seldom seen in Florida.

At one time Sphaeropsis knot was said to occur in all citrus-growing areas of Florida (272), but now it is seldom encountered. In Jamaica, however, the disease has lost none of its destructiveness.

Often confused with Sphaeropsis knot in Florida are the many types of galls that on investigation prove to be nontransmissible (197). Sphaeropsis knot occurring below the bud union of trees on rough lemon can be mistaken for woody gall, a virus disease not yet known in the state (189).

Sphaeropsis knot occurs on young and old branches of West Indian lime, sweet and sour orange, Ortanique, grapefruit, and rough lemon, eventuating in tree decline and death. Knots develop at or between the nodes. At first they consist of slight swellings covered by smooth, light-colored bark; later they increase in diameter from ⅜ to 3 in. and are covered by bark that is darkened and cracked (Fig. 70). Old knots are devoid of bark and are deeply furrowed, hard, blackish, and difficult to remove because of their broad basal attachments. Knots are usually solitary but at times several may coalesce. Buds proliferate over the surface and may sprout to produce witches' brooms. Attacks on the rough-lemon rootstock portions of trunks in Jamaica lead to a decline and ultimate death of young trees (26).

The cause of Sphaeropsis knot is *Sphaeropsis* (=*Haplosporella* of some authors) *tumefaciens*

Hedges, a fungus that ramifies in the cortex and at times in the wood of knots and produces masses of black mycelium. The pycnidial fruiting bodies are not encountered in nature but have been seen on artificially inoculated plants in the greenhouse (154).

Under Florida conditions, where the disease is mild in its development and restricted in its ability

FIG. 70. *SPHAEROPSIS KNOT.* Tumors on young branch and twigs of acid lime in Jamaica. (Courtesy of A. N. Naylor.)

to spread, adequate control has been obtained by removing and burning affected limbs.

SPHAEROSTILBE. Genus of the ascomycetous fungi. *A. aurantiicola* (B. & B.) Petch overgrows dead scale insects. See **Entomogenous Fungi.**

SPREADING DECLINE. A decline of trees caused by the burrowing nematode.

Within the past four decades, an eelworm invisible to the naked eye has destroyed over 10,000 acres of citrus, and for the past 25 years has occupied the full-time attention of pathologists and regulatory officials. Losses in production and expenses of control have made spreading decline the most costly citrus disease in Florida's history.

Plants affected

The burrowing nematode attacks indiscriminately all of the common rootstocks: rough lemon, sweet and sour orange, grapefruit, Cleopatra mandarin, Rangpur lime, and trifoliate orange. Of more than 1,400 different types of citrus that have been tested (113), only 5 promise relief if planted in or near cleared and fumigated areas. The Milam, a rough-lemon hybrid, derives its value from preventing the development of nematode eggs in the rootlet cortex and in this way reducing ultimately the nematode population in the soil. The Ridge pineapple, a midseason sweet-orange variety, is considered resistant to the nematode because rootlet growth is not retarded and because nematodes gradually disappear from the root system. The Estes, a type indistinguishable from rough lemon, is considered tolerant because rootlet regeneration is more rapid than root destruction by nematodes. The Carrizo citrange, a hybrid of navel orange and trifoliate orange, is considered tolerant, but it lacks adequate field testing. The Algerian Navel sweet orange is resistant to rootlet damage but it, too, requires further testing under field conditions.

There is no evidence that scion varieties influence the susceptibility of rootstocks to the burrowing nematode.

In addition to citrus, 225 other plants are hosts of the burrowing nematode (258). Among the more commonly encountered weed and cultivated species are acacia, allamanda, avocado, bamboo, banana, bean, beggarweed, camphor tree, casuarina, dogwood, eucalyptus, fig, gardenia, grasses (among them, nut, crab, Bermuda, maidencane, sugarcane, sandspur, and Pensacola bahia), guava, persimmon, various pines, strawberry, tomato, and watermelon. The presence of these hosts has a direct bearing on the dissemination and control of spreading decline in that these plants, if harboring the nematode, may be the source of inoculation for citrus, and in that these hosts when growing as weeds may provide food for the nematode in areas that have been inadequately fumigated. If ornamentals or crop plants are to be grown adjacent to citrus groves or in infested soil, their selection should be limited to those tested and known to be nonhosts of the burrowing nematode (258).

Though the nematode is present in many parts of the world, it has not been reported destructive to citrus except in Florida.

History and distribution

Spreading decline was first observed in the late 1920s in the Winter Haven area. By 1936, the trouble had spread to groves in other parts of Polk County, and by 1956, it was known in 1,053 citrus groves, 130 citrus nurseries, 179 ornamental nurseries, and 109 dooryard locations in 32 counties. The disease takes it greatest toll in sandhill locations on the Ridge. In flatwoods areas spreading decline is rarely encountered.

Up to December 1970, 2,332 groves have been found infected and 11,341 acres have been pulled and treated (257).

Importance

Losses from spreading decline are experienced by growers, by ornamental and citrus nurserymen, and—because of the expenses of control measures and research—by the state.

Growers suffer because declining trees do not yield sufficient fruit to pay the cost of production. Figure 72 shows the effect on production over a 10-year period and substantiates grower experience that cropping is reduced 40–80 per cent.

Nurserymen have been profoundly affected by the need to treat infested stock with hot water, by quarantine restrictions, and by regulations limiting future nurseries to sites found free of the nematode.

The state has spent millions of dollars in research, in surveys for the distribution of spreading decline, in pulling infested trees, in fumigating land preparatory to replanting, and in establishing and maintaining buffer zones to keep the nematode out of noninfested groves. To date, approximately 800,000 trees have been pushed and the soil treated with nematicides. As of 1966, 211 miles of buffer zones, 16–100 feet wide, have been installed around infested properties in attempts to protect 1,099 commercial grove properties (258). As of December 1970, 971 commercial groves remain to be pulled and treated and 95 groves remain to be ringed by buffer zones (257).

Symptoms

Symptoms produced by the burrowing nematode in the tops of trees are those of a generalized decline and are easy to confuse with declines due to other causes such as water damage, meteorological disturbances, mite infestations, mineral deficiencies, virus diseases, and poor cultural practices. As with many types of decline, trees affected by the burrowing nematode exhibit sparse foliage, small, pallid leaves, erratic flowering, dieback of twigs, and fruits that are reduced in size and number. Affected trees do not die but neither can they be made to recover.

The two characteristics that enable ready identification of spreading decline are the sharp margin (Fig. 71A) between healthy and diseased portions of the grove (all trees behind the margin being about equally affected) and the inexorable forward movement of the disease front, involving about 2 trees per row per year. Trees recently infected take about 2 years to decline. Usually the first 2 rows of trees on the healthy side of the margin are already parasitized.

Symptoms underground include a scarcity of feeder roots at depths below 10 in. (above 10 in., feeder roots are often more plentiful than in healthy trees). On the average, an affected tree has half the normal complement of feeder roots and so is unable to absorb sufficient water and nutrients to prevent early wilting and insure normal growth.

Inspection of feeder roots shows lesions on the exterior surface (Fig. 71D) and cavities in the cortex (Fig. 71C). Confirmatory evidence of spreading decline is the detection of the causal nematode in the roots.

Cause

In 1953, the cause of spreading decline was reported (320) to be the burrowing nematode, *Radopholus similis* (Cobb) Thorne, 1949, synonymous with *Tylenchus similis* Cobb, 1893, *Tylenchus acutocaudatus* Zimmermann, 1898, *Tylenchus biformis* Cobb, 1909, *Anguillulina similis* (Cobb, 1893) Goodey, 1932, and *Rotylenchus similis* (Cobb, 1893) Filipjev, 1936. Females taken from citrus are 0.65–0.80 mm long by 20–24 microns in diameter, with vulva located at 54–59 per cent of the distance to the tail, with head rather flattened and only slightly offset from the body by a constriction, and with a heavy (ca. 17 microns long) stylet subtended by 3 distinct basal knobs (Fig. 71B). Males are 0.5–0.65 mm long, more slender than females, with head spherical and sharply set off from the body, with stylet weak, short, and basal knobs difficult to see, and with a flap near the tail end (321).

The life cycle is completed in 19–21 days at 75–78°F; thus, under favorable conditions, populations increase rapidly. The period of longevity is not known, but survival of nematodes in rootlet-

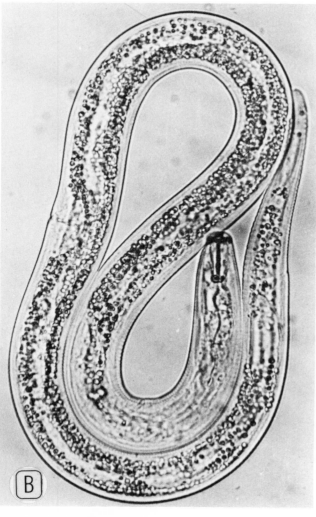

FIG. 71. *SPREADING DECLINE.* A. Left, healthy, and right, diseased sides of margin. B. *Radopholus similis* (female). C. Cavities and nematodes in feeder root. D. Lesions on surface of feeder roots. E. Nematodes within feeder root. (Photograph A by R. F. Suit, B by R. P. Esser, C and D by E. P. DuCharme, E by H. W. Ford.) (For Figs. D and E see next page.)

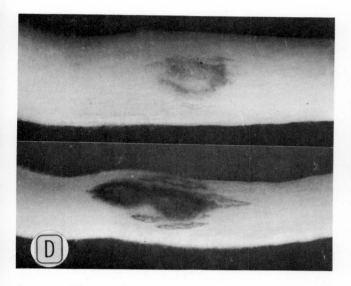

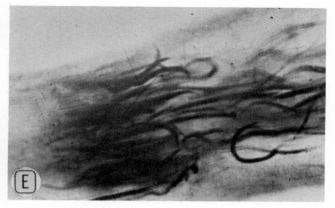

free sand has not been observed beyond 6 months.

The top 6 in. of soil are usually free of nematodes due to adverse temperature relations. The largest number of nematodes is found at depths of 1–6 feet, though some may be found down to 12 feet. The burrowing nematode attacks young, healthy, unsuberized rootlet tips and feeds on the parenchymatous tissues of the cortex and stele, forming cavities. Eggs laid inside a cavity hatch into larvae that remain to feed in the same location (Fig. 71E) for as long as food is available. Once the food supply is exhausted or secondary fungi and bacteria decompose the host tissue, nematodes migrate to fresh rootlets. The migration is reflected aboveground in the further inroads made by spreading decline.

Two races of *R. similis* have been encountered in Florida (321). Though indistinguishable by inspection, they differ in their pathogenicity to certain groups of plants. One causes decline in citrus, avocado, Barbados cherry, ginger lily, and banana, whereas the other (the so-called banana strain) does not attack citrus.

The burrowing nematode is spread to new locations by the movement of infested nursery stock (whether citrus or ornamental) and by the transport of implements or of sand and clay fill from infested properties. Once established in an area, nematodes are said to extend their range, principally by root-to-root contact (even across highways) but also by cross cultivation from an infested to a noninfested portion of the grove and by run-

off of rain or irrigation water from infested areas (258).

Control

The objectives set by the Florida Department of Agriculture and Consumer Services, which is the agency responsible for the control of spreading decline, are twofold: containment of the nematode to areas already infested, and ultimate eradication of the nematode from the state. In furtherance of the first objective, 8 types of inspections are conducted routinely: detection inspections, both initial (on properties that heretofore have not been sampled) and reinspection (on properties previously not found to be infested); confirmatory inspections (to extend sampling where only a few nematodes have been found); delimiting inspections (to determine the exact area of infestation preparatory to fumigation); margin inspections (to determine the success of fumigation treatments in buffer zones around the spreading decline periphery); emplaced buffer inspections (to determine the effectiveness of fumigated barriers); replant inspections (to ascertain the nematode status of trees reset in pulled and treated groves); real estate inspections (to determine, for the intended buyer of a grove, the presence or absence of the nematode); and nursery inspections (to insure the production of nematode-free nursery stock).

Once an infestation has been identified, two alternatives are available to the grower. He may elect to pull affected trees plus the two healthy-looking trees in each row bordering the decline area and to fumigate the soil with DD (dichloropropane-dichloropropene) at the rate of 60–100 gal./acre, or with Telone, 45 gal./acre, injected in rows 18 in. apart and to a depth of 12 in. The area treated must be kept free of vegetation for at least 6 months. The alternative is to ring the infested area with a fumigated barrier at least 32 feet wide. The fumigant presently used is ethylene dibromide (W-85) at 50 gal./acre for the initial application and 25 gal./acre for subsequent applications every 6 months. For the suppression of weed growth after

A comparison of production over a ten-year period between two ten-acre citrus groves of identical variety and type, with the exception that one grove was planted with healthy nursery stock and the other planted with burrowing nematode infested nursery stock. Both groves were planted during the same season on land free of burrowing nematode.[1]

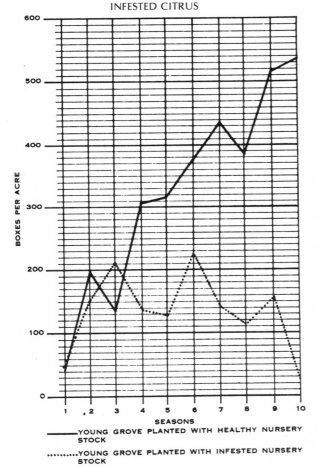

TEN-SEASON PRODUCTION COMPARISON
HEALTHY CITRUS vs. BURROWING NEMATODE
INFESTED CITRUS

——— YOUNG GROVE PLANTED WITH HEALTHY NURSERY STOCK

·········· YOUNG GROVE PLANTED WITH INFESTED NURSERY STOCK

1. All figures furnished by production foremen from the two major growers concerned. Figures compiled by Charles Poucher, chief of the Division of Plant Industry Special Programs Section.

FIG. 72. *SPREADING DECLINE.* The relation of spreading decline to crop production over a 10-year period. Age of trees at Season 1 was 5 years. (Reproduced from *Burrowing Nematode in Citrus,* Florida Dept. Agr., Div. Plant Ind. Bull. 7, 1967.)

fumigation, Diuron is applied at the rate of 40 lb./acre.

Following a waiting period of 1 year after the push-and-treat operation, the area can be reset to citrus, using nematode-resistant rootstocks. Great care must be taken not to replant with nematode-infested nursery stock.

Latest regulations governing matters of control and the obligations of nurserymen in providing nematode-free stock are available from the Division of Plant Industry, Florida Department of Agriculture and Consumer Services, Gainesville, Florida.

STAR MELANOSE. A stellate-shaped cracking of melanose pustules on leaves and an intensification of blemishes on fruit due to late applications of copper-containing fungicides.

Melanose pustules on leaves explode when sprayed with copper (Fig. 73A), and rind blemishes due to melanose, wind scar, and russet are darkened. The intensification of rind blemishes (Fig. 73B) may result in a lowering of grade in the packing-house.

Star melanose is not to be confused with other star-shaped blemishes on the rind, e.g., algal spot (Fig. 1C) which is a black growth on overripe fruit that can be removed by brushing; the depressed "star-shaped spot," thought to be due to an inherent weakness of the rind (93); and the stellate Atichia fungus which attaches to bodies of purple scale (80).

Star melanose can be prevented by correct timing of copper sprays. For proper control of melanose, applications should be made before pustules appear, not after. Copper sprayed on melanose-affected trees as late as the first week in June may not be completely effective and considerable star melanose may still appear (317). Whether summer applications, as for the control of greasy spot, may continue to change melanose pustules into star melanose is not a matter of record. In the absence of information on this point, it seems preferable to use oil (1 per cent FC 435-66) in the July greasy-spot control program, especially if the rind is much marked by melanose, wind scar, or russet, and fruit is intended for the fresh market.

STEM-END PEEL INJURY. See **Stem-End Rind Breakdown.**

STEM-END RIND BREAKDOWN. A postharvest rind collapse of citrus fruits. *SYN.*—"Serb," brown stem, burnt stem, stem-end peel injury, aging, and gas burn (erroneously).

Stem-end rind breakdown (Fig. 74) is the most common type of peel injury affecting oranges in storage. A similar but perhaps unrelated condition is seen at times in tangerines, Temples, tangelos,

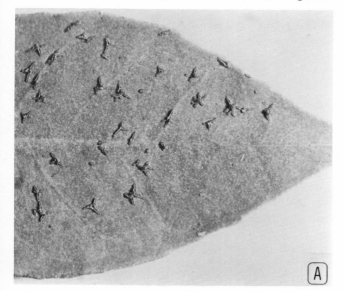

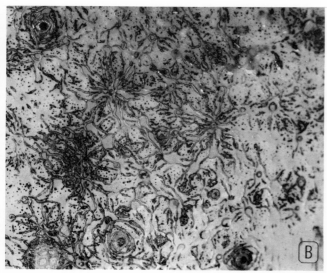

FIG. 73. *STAR MELANOSE.* A. Star-shaped cracking of melanose pustules on leaf. B. Intensification of melanose blemish on fruit. (Photograph B by J. F. L. Childs.)

and grapefruits. The disorder generally appears 2–7 days following harvest.

The rind breakdown starts at the stem end, but the button is usually separated from the involved area by a narrow ring of intact tissue. With time, affected portions turn brown or black. Lesions expand as long as fruits are kept under drying conditions.

Stem-end rind breakdown results from excessive dehydration. It can be prevented by keeping fruits in the shade after picking, by covering loads in transit with canvas, by shortening the interval between harvesting and waxing, and by maintaining high humidities (88–97 per cent) in the degreening room (222).

STEM-END ROT. See **Diplodia Stem-End Rot, Phomopsis Stem-End Rot, Black Rot.**

STEM-PITTING DISEASE. See **Tristeza.**

STIGMANOSE. Round, brown, glazed spots, 1/8–3/8 in. in diameter, found occasionally on the upper surfaces of leaves; thought to result from the feeding of pumpkin bugs (305).

STING NEMATODE. See **Belonolaimus, Nematodes.**

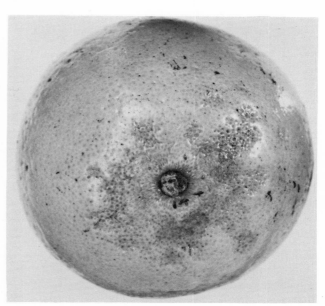

FIG. 74. *STEM-END RIND BREAKDOWN.* The areas of collapsed rind range in color from brown to black. (Photograph by A. A. McCornack.)

STRANGLE WEED. See Dodder.

STUBBY-ROOT NEMATODE. *Trichodorus christiei* Allen has been shown to be pathogenic to citrus. See **Nematodes.**

STYLAR-END BREAKDOWN. A rind collapse of limes, limequats, and lemons. *SYN.*—stylar-end rot.

Large losses to the Tahiti lime crop are occasioned some years by a rind collapse known as stylar-end breakdown (Fig. 75). Other varieties also affected include Key limes, limequats, and

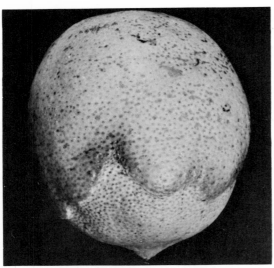

FIG. 75. *STYLAR-END BREAKDOWN.* Tahiti lime exhibiting a rind breakdown that at first is water-soaked and then turns leathery and brown.

lemons. The disorder is more prevalent in large and mature fruits, especially when harvested in late summer and early fall (79).

The breakdown begins as a water-soaked but firm area at or near the stylar end. The lesion enlarges rapidly—either in fruits on the tree, in storage, or in transit—until as much as one-third to one-half of the rind is involved. With time, affected areas dry out, settle, turn brownish, and become demarcated from healthy rind by a dark ring.

The cause of stylar-end breakdown is not known. Microorganisms are absent during early stages of the disorder, but secondary fungi such as *Penicillium*, *Colletotrichum*, *Diplodia*, *Gloeosporium*, and *Geotrichum* are present in later stages. The

trouble is believed to be related to moisture stress during maturation of fruits on the tree. It has been suggested that incidence in the field can be reduced by frequent irrigation. Losses at harvest time may be minimized by picking fruits before rind breakdown appears. Bruising of fruit during handling increases markedly the amount of the trouble in storage (86, 150, 283). Limes held at 100°F develop significantly more stylar-end breakdown than those held at 40–90°F (150).

STYLAR-END ROT. See Stylar-End Breakdown.

STYLAR-END RUSSET. A superficial rind blemish usually restricted to the blossom-end half of the fruit.

Stylar-end russet is an occasionally encountered blemish, resembling the one caused by citrus rust mites but differing in being restricted to areas of the rind near the stylar end and in consisting of a network of fine corky lines. At one time, the cause was suspected to be the citrus red mite (*Metatetranychus citri*) (259). However, recent investigations show that mites are not involved. The cause remains undetermined.

SULFIDE INJURY. See Water Damage.

SUNSCALD. Damage to bark and fruit from the combined action of excessive light and heat.

Sudden removal of normal shade subjects bark to scalding by the sun. Serious damage to limbs and trunks may follow hatracking or rapid defoliation brought on by diseases, insects, drought, cold, or waterlogging. Injured bark allows the entry of wood-rotting fungi and insects that, in turn, cause further destruction. The sunscald that follows topworking or hatracking can be avoided by whitewashing the bark with a mixture of 5–10 lb. hydrated lime and 1 lb. zinc sulfate with sufficient water for application by brushing or spraying.

Fruits are also subject to sunscald, especially after applications of sulfur during hot weather. Certain varieties such as the Murcott, the Temple, and the King orange are prone to sunburn even when not sprayed. In mild cases, sunscalding causes a pitting of the rind; in severe cases, it produces hard, gray-black scabs up to 1.5 in. in diameter. Fruits sunburned while small may become distorted at maturity.

SYMPTOMLESS INFECTIONS. Fungal invasion of host tissue without the production of symptoms or the appearance of fungal fruiting bodies until tissue resistance has been lowered or destroyed by other agents. SYN.—latent or quiescent infections.

When isolations are made from apparently sound citrus tissues, certain fungi (e.g., *Colletotrichum gloeosporioides*, *Alternaria* sp., and *Guignardia* sp.) emerge in culture (1, 13, 223). The colonies do not result from surface contamination (which can be guarded against by sterilization procedures) but from mycelium present within the host tissue. As is well known from studies on other plants, fungi may be present in host tissue without causing symptoms (129). When, however, the normal defenses of the host are diminished, as happens when tissues are weakened or killed by other agents, these fungi either establish visible infections or begin the saprogenic stage of their life cycle, producing fruiting bodies. Such fungi, therefore, cannot be considered the cause of disease even though they are present in diseased tissue. This understanding of the behavior of these fungi accounts for the reason why, for example, *Colletotrichum gloeosporioides* should no longer be considered the cause of dieback or anthracnose.

TAR SPOT. An occasionally encountered foliar, twig, and fruit spotting resembling greasy spot. SYN.—spackle.

Tar spot has been described as a lesion on leaves and twigs that resembles greasy spot but that differs in manifesting a mahogany-red ring inside the lesioned area. Fruits may also be affected, in which case they show greasy-spot-like blemishes on the rind (Fig. 76). The cause is given as *Cercospora gigantea* Fisher (106).

Because of its infrequent occurrence, tar spot warrants little concern. Control measures have been reported (107).

TATTER LEAF–CITRANGE STUNT. A virus disease complex affecting certain citranges and limes but in Florida found only as a latent infection in Meyer lemon trees.

A testing of 24 randomly selected Meyer lemon trees in Florida revealed 59 per cent to contain the tatter leaf–citrange stunt virus complex, usually admixed with tristeza virus (121, 122). Infected trees showed no symptoms, but inoculated seedlings of *Citrus excelsa*, citremon, and several varieties of citrange showed a mottling and distortion of the foliage (121, 122, 345, 346). These symptoms (Fig. 77A) usually appear 4–8 weeks after inoculation. Several years after inoculation, trees of Satsuma mandarin on Troyer citrange in California developed a vertical corrugation of the stock portion of the trunk, a groove at the bud union, and a decline of the top (41). Similar vascular abnormalities have since been found in Florida (122) in experimentally inoculated young trees of sweet orange on trifoliate orange, Troyer, Carrizo, and Rusk (Fig. 77B).

The virus complex is transmissible by grafting and sap inoculation and produces symptoms in the aforementioned species as well as in cowpea and certain other herbaceous plants (125).

Tatter leaf–citrange stunt virus complex is of potential importance to Florida because of the increasing use of citranges and trifoliate oranges as rootstocks. To date, however, there is no evidence that the virus is present in commercial scion varieties other than Meyer lemon (345) or that the complex is spread by vectors from infected Meyer trees to nearby trees of other varieties (121). The indexing of budwood-source trees on Rusk citrange and *C. excelsa* provides a rapid means of determining whether the virus complex is present. As precautions, only budwood free of the complex should be used on susceptible stocks, and only virus-free Meyer trees should be propagated.

TEARS. See **Rio Grande Gummosis.**

THREAD BLIGHT. A disease of minor importance on citrus; damage is limited mostly to a blemishing of the rind.

Thread blight is a common and often destructive disease of many plants in the tropics. In Florida, it is encountered rarely on citrus and causes little more than a rind blemish.

The causal fungus is *Corticium stevensii* Burt (356), sometimes given as *C. koleroga* (Cooke) Hoehn. (374). It produces black, string-like rhizomorphs on leaves, twigs, and fruits. The strands terminate in black sclerotial bodies of various sizes and shapes. Leaves may be killed and matted together and may remain attached to trees by the rhizomorphic strands.

Control measures have not been required, but if thread blight ever became serious, it could be controlled by spraying with the same copper fungicides that are recommended against melanose.

TILLANDSIA. Genus of bromeliaceous air plants. One species, *T. usneoides* L., is known commonly as **Spanish Moss** (which see).

TRICHODORUS. Genus of the phytophagous nematodes. *T. christiei* Allen (one of the stubby-root nematodes) has been proven to be pathogenic to citrus, but its economic effects remain to be assessed. See **Nematodes.**

TRISTEZA. A virus disease causing a decline of certain stock-scion combinations and even of some varieties when grown as seedlings. *SYN.*—quick decline, stem-pitting disease, bud-union decline disease.

During the past 30 years, tristeza has destroyed hundreds of thousands of trees in California and over 20 million trees in Argentina and Brazil.

Tristeza was discovered in Florida in 1951 (134). In retrospect, however, the virus may well have been introduced as early as 1908 when tristeza-infected Meyer lemon trees were imported from China.

Dissemination of tristeza in Florida has generally been slow in comparison with spread in other countries. The initial survey for tristeza in Florida showed that in 1953 approximately 500 groves in 27 counties contained some declining trees (72). Between 1953 and 1961, tristeza virus was detected in 10 per cent of 2,446 healthy appearing trees in various parts of the state. Of 689 trees tested that were free of tristeza virus prior to 1958, 7.2 per cent had become infected by 1961. The area of most rapid spread was Orange County; here, 38 per cent of trees known to be free of the virus prior to 1958 had become infected. The area with the lowest rate of spread was the east coast (238).

Since 1961, severe losses from tristeza have occurred in such localities as Elfers in Pinellas County, Auburndale and Polk City in Polk County, and the Lake Apopka area in Orange County (28, 192). There is now evidence that the virus is spreading on the east coast—an area where the majority of trees are on the susceptible sour-orange rootstock (238).

In the past, losses from tristeza have not been great in Florida, perhaps because prevalent strains of the causal virus are not virulent (many trees

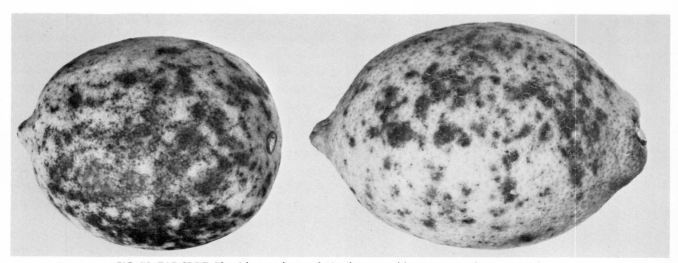

FIG. 76. *TAR SPOT*. Blemishes on lemon fruits that resemble greasy spot lesions on foliage.

FIG. 77. *TATTER LEAF–CITRANGE STUNT.* A. Symptoms of infection on leaf of Rusk citrange. B. Groove at bud union in a decorticated trunk of sweet-orange tree on Rusk citrange. Extent of bud-union abnormality indicated by the separation of stock from scion. (Photographs by S. M. Garnsey.)

known to have been infected 6–10 years still show no symptoms) and because spread by aphid vectors occurs only during occasional years when populations are high (296). But as more and more trees in the state become infected, including those on tolerant stocks, the rate of spread can be expected to accelerate.

Plants affected

In Florida, tristeza is a threat only to trees of sweet orange, grapefruit, mandarin, and various of the specialty fruits when grown on sour-orange rootstocks. Elsewhere in the world, tristeza may also affect grapefruit (249) and some varieties of sweet orange (231) irrespective of rootstocks, causing a variant type of tristeza known as stem-pitting disease. Though many grapefruit trees in Florida carry the tristeza virus, they do not show the symptoms of stem-pitting disease (189, 206).

In addition to sour orange, many other varieties, mostly of minor importance in Florida, are susceptible when used as rootstocks (23, 231). Some of these are alemow *(Citrus macrophylla),* shaddock, dense-leaved box orange *(Severinia buxifolia),* certain tangelos, and certain limes and lemons (both sweet and acid). In California and Brazil, but so far not in Florida, Troyer citrange has been found to be susceptible (42). Although lemon trees on alemow are not usually infected in the field, they may become infected and decline if aphid vectors inject tristeza virus into alemow seedlings in the nursery or if they inject the virus into root sprouts of trees in the field (40).

Certain varieties show vein-clearing or stem-pitting symptoms irrespective of rootstocks or even when grown as seedlings: Key and Tahiti lime, Eustis limequat, Nagami kumquat, *Citrus hystrix, Pamburus missionis, Afraegle paniculata,* and *Aeglopsis chevalieri* (180).

The percentage of infection in Florida varies according to the scion variety; thus, in one survey, the rate in Temple and sweet-orange trees was found to be 20–23 per cent, in mandarins 9 per cent, and in grapefruit 2 per cent (238). These percentages, based on trees that became inoculated by aphids, indicate perhaps the preference shown by aphids for these varieties.

Elsewhere in the world, certain scion varieties when budded on infected but tolerant rootstocks have been found affected by tristeza. Thus, in Brazil, decline has been seen in trees of Key lime, grapefruit, Mediterranean sweet orange, and true lemon when grown on tolerant stocks (231).

Rootstock varieties generally considered tolerant to tristeza virus are rough lemon (including Estes and Milam),[1] *Citrus volkameriana,* sweet orange, Cleopatra mandarin, Rangpur lime, trifoliate orange, and citrumelo (202, 231, 289).

According to some authorities, tristeza virus is a complex of several components. One component, seedling yellows virus, also affects seedlings of sour orange, grapefruit, and acid lemon, causing chlorosis and stunting under greenhouse conditions (344). In Florida, field transmission of seedling yellows virus to these susceptible species is not known to occur.

Symptoms

Affected trees show many symptoms but diagnosis is often difficult because most of the symptoms lack specificity. Leaves exhibit bronzing and various deficiency symptoms, foliage wilts, leaves drop, and twigs, branches, and roots die back. Trees usually linger in an unproductive state for many years (Fig. 78) but occasionally die within a few months of showing initial symptoms, in which case they retain their leaves in a parched condition (Fig. 79). But these symptoms are much the same as those produced by any root-pruning agency (e.g., trunk girdling, water damage, nematodes, and fungi) and so are not reliable for diagnosing tristeza.

The following procedure is useful in diagnosing tristeza-affected trees.

(a) Tristeza may be involved if top symptoms appear only in trees on sour-orange stocks. If adjacent trees on other rootstocks (e.g., rough lemon, Cleopatra mandarin, or sweet orange) are similarly affected, the decline is probably due to some other agency such as blight, weevil damage, nematodes, or unfavorable water relations.

(b) If decline appears only in trees on sour orange, proceed to an examination of the bud union. If infection by tristeza virus occurred years

ago, the scion portion of the trunk will be bulged out over the stock portion (Fig. 78). If infection by aphids occurred recently, the bulge symptom may not have had time to develop.

(c) Whether or not bulge appears, proceed to the removal of inch-square patches of bark across the bud union, and examine the inner faces with a hand lens for the presence of "honeycombing" (Fig. 80)—a massing of many fine holes in the rootstock portion just below the bud union. Although honeycombing may result from other causes (191), it has been found to be indicative of tristeza in 90 per cent of trees sampled in Florida (72).

(d) Confirmation of these field symptoms is obtained in the greenhouse by inserting buds from suspect trees into Key lime indicator seedlings. If tristeza virus is present, vein-clearing (Fig. 81A) and leaf-cupping symptoms usually appear in from 30

FIG. 78. *TRISTEZA.* Twenty-five-year-old sweet-orange tree on sour-orange rootstock, showing chronic type of decline usually encountered in Florida.

1. Seeds from 33 rough-lemon source trees in Florida were grown off and allowed to become infected with the virulent strains of tristeza in Argentina. All 33 sources proved tolerant (202).

to 50 days in the lime leaves that unfold after grafting. In time, stem-pitting symptoms (Fig. 81B) develop on the woody cylinder of the lime. In Key lime, Tahiti lime, and limequat trees in the field, vein-clearing and stem-pitting symptoms may be used directly as evidence of infection.

Tristeza may also be diagnosed in susceptible stock-scion combinations by detecting necrosis in phloem elements and adjacent parenchyma-like cells (290, 291). Electron microscopy of partially

FIG. 79. *TRISTEZA.* Acute collapse, occurring at times within months of the first appearance of symptoms.

purified virus in the sap of plants (14) or serology (250) show promise for identifying the virus.

For all practical purposes, indexing is an adequate procedure for determining the presence of tristeza virus, but for assaying virulence or for analyzing various components of the tristeza virus complex, other procedures are required (344).

In Florida, trees of susceptible combinations may carry the tristeza virus for many years without showing symptoms (69). Some authorities believe such trees may suddenly decline once they encounter conditions of stress as might be brought on by drought or waterlogging (238).

Cause

The cause of tristeza is a phloem-localized virus. Various theories prevail with respect to the complexity of the virus. The major hypotheses are: (a) that a single virus is involved and that its consequences to the tree vary with the pathogenicity of various strains (206, 314) and (b) that the virus is a complex consisting of such components as a

seedling yellows factor and a stem-pitting factor (116, 219, 344).

Tristeza particles are flexuous strands 10–12 millimicrons wide by approximately 2,000 millimicrons long (15, 168, 261).

Under field conditions, the virus or virus complex is transmitted by budding and by certain species of aphids. Vectors that occur in Florida are *Aphis gossypii, A. spiraecola,* and *Toxoptera aurantii* (240). Elsewhere, other vectors are *T. citricida, A. craccivora, Dactynotus jaceae,* and *Myzus persicae.* An illustrated key for identifying the various species of aphids found on citrus has been published by Stroyan (312).

Tristeza virus has also been transmitted through *Cuscuta americana* (207) and *C. subinclusa* (353). Seed transmission has not been demonstrated in numerous attempts (218).

Control

Tristeza can be avoided to only a limited extent through the use of virus-free budwood, because aphid vectors may subsequently infect trees in the field. Little reduction in the rate of spread can be expected by attempting to control the vector since insecticides usually require 20 minutes to effect kill whereas aphids can inoculate trees within seconds.

A simple method of control is to use as rootstocks only those varieties listed as tolerant under *Plants affected.* The selection of tristeza-tolerant rootstocks should be tempered, however, by considerations of their susceptibility to other virus diseases such as exocortis and xyloporosis.

Despite the threat of tristeza, over 30 per cent of trees being grown in Florida nurseries today are on sour orange, which attests to the popularity of this variety. Much effort has gone into the finding of sour oranges that might tolerate tristeza virus. Several clones show promise in Australia (313) and in Florida (142), but none can be recommended until disease relations and horticultural characteristics have been more fully investigated.

Attempts are also being made to immunize trees by preinfecting them in the nursery with weak strains of the tristeza virus in the hope that they will be protected from more severe strains when later inoculated by aphids in the field. Promising results from preimmunization have been reported from Brazil (232), Australia (118), and California (347, 348), but the method cannot be recommended until more is known of the duration of protection and of the mixture of weak strains required to protect trees from all virulent strains transmitted in the field.

Because tristeza virus is not seed-borne, virus-free budwood can be obtained from seedlings provided they have not been inoculated by aphids. For scion varieties that do not reproduce true to type, virus-free propagative material has been obtained from infected plants by growing them in a heat chamber at 90–110°F for 40 days and then propagating the tips of new shoots formed in the chamber (134). Another technique proposed for deriving virus-free propagative material from monoembryonic varieties is to culture nucellar tissue on agar in order to promote the development of apomictic seedlings (24).

Diseased trees cannot be cured, and their condition cannot be ameliorated by supplemental irrigation or fertilization. Some expedients may be used to improve tree performance, such as inarching and scion rooting (210), but these practices are uneconomical and cannot be recommended except for the salvation of specimen trees. Once trees decline to the point of unprofitability, they should be removed and replaced with trees on tolerant rootstocks.

TWIG DIEBACK OF ROBINSON TANGERINE. See **Robinson Dieback.**

TYLENCHULUS. Genus of the nematodes. *T. semipenetrans* Cobb, 1913, is the cause of **Slow Decline** (which see).

USTULINA. Genus of the ascomycetous fungi. *U. vulgaris* Tul. has been reported as a secondary invader in foot rot lesions (93).

VERMICELLA. See **Dodder.**

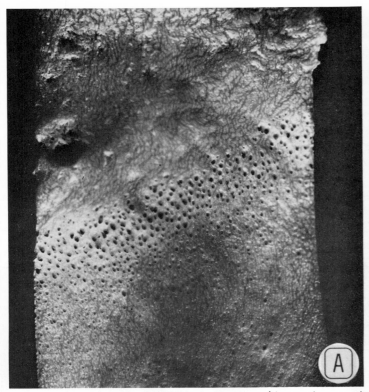 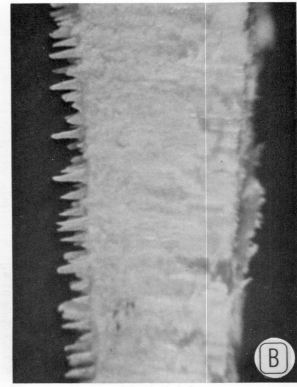

FIG. 80. *TRISTEZA.* A. Honeycombing or inverse pitting in the inner face of the bark below the bud union. B. Bristles on the face of the wood below the bud union. These bristles fit into the holes shown in A. Both views much enlarged, B more than A. (Photograph A by W. C. Price.)

VERRUCOSIS. See **Scab.**

VERTICILLIUM. Genus of the deuteromycetous fungi. *V. cinneamomeum* Petch overgrows dead scale insects. See **Entomogenous Fungi.**

WART. An apparently genetic abnormality, leading to scabrous outgrowths on the trunks of Valencia trees.

Wart is a rarely encountered abnormality of young Valencia orange trees. Eruptive galls ¼–½ in. in diameter cover the trunk from bud union to scaffold branches (Fig. 82A). Surfaces of galls are cracked and rough. On removal of the bark, the woody cylinder shows brown, cone-shaped cavities

extending into the wood from ¼ to 1 in. (Fig. 82B).

The trouble has not proved to be bud-transmissible, though in greenhouse tests a twisting and chlorosis of the leaves, similar to that found in crinkle-scurf, develop in the foliage of sprouts from affected buds. Seeds from affected trees give rise to seedlings with leaves having undulating margins (197).

WATER DAMAGE. The most prevalent cause of citrus decline and loss of production in Florida.

Seventy-eight per cent of Florida's citrus trees are located on flatwoods soils.[1] The problems pecu-

1. Estimated from 1969 inventory of commercial trees in *Florida Agricultural Statistics*, December 1, 1970. Florida Crop and Livestock Reporting Service, Orlando, Florida.

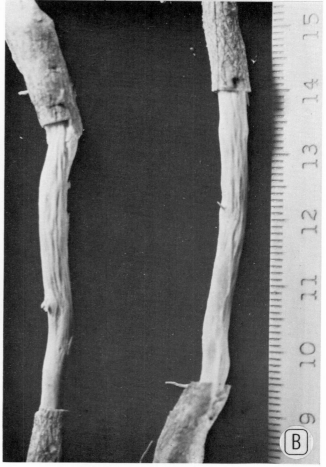

FIG. 81. *TRISTEZA.* A. Vein clearing in leaf of tristeza-virus inoculated Key lime. B. Stem pitting in wood of inoculated Key lime. (Photograph A by W. C. Price.)

liar to flatwoods areas, consequently, have a decided impact on the state's total production of citrus.

The major depressant to production in flatwoods areas is poor drainage along with associated problems such as root rot, foot rot, and sulfide damage. The water relations of trees and the management of flatwoods soils are complex subjects, and they fall outside the scope of this publication. The following discussion is limited to the diagnosis of declines brought about by chronic and acute types of water damage.

Chronic water damage may be expected in groves located near bodies of water or on shallow soils underlaid by impervious hardpan or rock. Where underground water accumulates and stagnates for prolonged periods, roots die from the toxic products associated with a lack of oxygen. Affected trees exhibit the same type of top decline that is associated with root reduction regardless of cause.

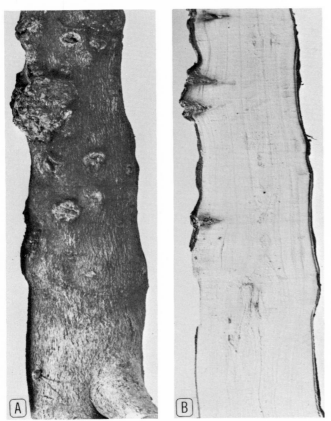

FIG. 82. *WART.* A. Rough-surfaced galls on trunk of Valencia sweet-orange tree on rough-lemon rootstock. B. Longitudinal section through trunk showing cone-shaped pegs extending into woody cylinder.

Leaves are sparse, small, and chlorotic, showing patterns similar to those of nitrogen deficiency. Trees are stunted and yield poor crops. Distribution of declining trees within a grove correlates with areas having poor drainage or layers of impervious clays close to the soil surface. When affected trees are pulled, root systems are found to be truncated, resembling a broom. The cortex of root ends is easily sloughed off to reveal blackened woody cylinders. Isolations from affected tissues frequently yield *Diplodia natalensis,* a fungus once thought to be the primary cause of root rot but now regarded as a secondary invader of weakened tissue. The only functional roots remaining may be those near the crown of the tree. Should these, too, be killed, as happens during prolonged periods of drought, the top collapses and the tree dies. The combined effect of high water table and drought is commonly spoken of as water damage when in reality it should be referred to as water and drought damage.

Acute water damage or flooding injury occurs when soil surfaces are flooded with water that remains stagnant. Little damage results from flood waters that flow internally at a rate equivalent to 6 in. of drawdown per day (114). The effects on tops and roots are similar to those of chronic water damage as induced by high water tables, except that leaves wilt suddenly and drop. If defoliation is only partial, there is a possibility of tree recovery. The severity of symptoms depends on whether sulfides are produced during the time roots are deprived of oxygen. Sulfides can be detected by smelling a handful of the flooded subsoil and roots.

Chronic water damage can be controlled by lowering the water table with tile lines or drainage ditches and by irrigating during times of drought to keep the remaining roots alive. Acute water damage can be minimized by rapid removal of standing water. Some rootstocks (e.g., rough lemon, sour orange, grapefruit, and trifoliate orange) can tolerate waterlogging longer than can others (e.g., Cleopatra mandarin) (112).

Water damage is easier to avoid than to remedy. Drainage problems should be investigated before trees are planted. Low-lying areas can often be utilized provided surface and subsoil drainages assure a well-drained rhizosphere to a depth of at least 3 feet and provided drainage ditches and tiles are properly engineered and maintained.

WILT. See **Blight.**

WIND SCAR. Rind blemishes produced by actions of the wind. *SYN.*—wind burn.

Wind scar (Fig. 83) is the most prevalent rind blemish on Florida-grown citrus fruits. Through the agency of wind, fruits are rubbed together (producing a yellow to gray scurfing at points of con-

FIG. 83. *WIND SCAR.* Various rind blemishes produced by wind-rubbing of fruit against limbs and thorns.

tact that suggests thrips injury) or against branches, are scratched by thorns, or are blasted by sand. The rind is especially tender during the 3-week period following petal fall. Injuries are generally superficial, but when large areas of the peel are involved, wind scar brings about a reduction in packout.

Little can be done to reduce wind scar short of planting windbreaks. Once injuries are produced, they become intensified with each application of copper fungicide. Alternatives to copper sprays are discussed under **Star Melanose.**

WINTER CHLOROSIS. See **Yellow Vein.**

WITHER TIP. See **Lime Anthracnose, Anthracnose.**

WOE VINE. See **Cassytha.**

WOOD POCKET. See **Lime Blotch.**

WOOD ROT. A fungus-induced rotting of woody tissues. *SYN.*—heart rot, concentric canker.

The wood of citrus trees consists of sapwood and heartwood. Sapwood serves as a conductive system for water and dissolved nutrients whereas heartwood provides structural support. The consequences of wood rot depend on which of these tissues is attacked.

If, on the one hand, fungal invasion is restricted to the heartwood, only structural strength is impaired. Trunks of old trees are often hollowed out by heart rot but because the sapwood is not involved, tops continue to bear acceptable crops. The economic effects of heart rot are limited mostly to the breaking of limbs and trunks by wind. In Florida, several fungi are associated with heart rot, including *Daldinia concentrica* (Bolt ex Fr.) Ces. & DeNot., *Xylaria polymorpha* (Pers. ex Fr.) Grev., and *Ganoderma sessilis* Murrill (Fig. 84C).

If, on the other hand, wood-rotting fungi also attack the sapwood, then the flow of water and dissolved minerals is impeded and as a consequence the foliage turns chlorotic and wilts and the top declines. In Florida, the fungus known to attack both heartwood and sapwood, and possibly also the bark, is *Fomes applanatus* (Pers.) Wallr., the cause of concentric canker (56). The wood rot so-named breaks through to the surface of trunks and limbs, leading to a killing and settling of the bark in a more or less concentric pattern (Fig. 84A).

Wood-rotting fungi enter trees through injuries to the bark. Some gain access through roots damaged by cultivation (Fig. 84B), fungi, arthropods, rodents, and water. Others gain entrance through aboveground wounds that are caused by bark rots, limb splitting, sunscalding, freeze damage, mechanical injury, and careless pruning or topworking.

To the extent possible, damage to the bark should be avoided. If incurred, large lesions or

FIG. 84. *WOOD ROT.* A. Concentric-canker type of wood rot with involvement of the bark. B. Progression of wood rot as revealed by longitudinal section of the trunk. C. Fructifications of *Ganoderma sessilis,* indicative of rotted heartwood. (Photographs B and C by E. P. DuCharme.)

wounds should be treated by cutting away diseased tissue, by disinfesting with Arvenarius Carbolineum, and by protecting exposed surfaces with asphalt wound dressings (replenished periodically until wounds have healed). Pruning cuts should be made through live wood and flush with limbs or trunks so that callus formation will cover openings quickly. If breaks, cuts, or injuries in the bark are small (less than 1 in. in diameter), the natural defenses of the tree are usually adequate to check invading fungi.

WRINKLE-RIND. See **Rumple.**

WRINKLE SKIN. See **Creasing.**

XANTHOMONAS. Genus of the phytopathogenic bacteria. One species, *X. citri* (Hasse) Dowson, causes **Canker** (which see).

XYLARIA. Genus of the basidiomycetous fungi. One species, *X. polymorpha* (Pers. ex Fr.) Grev., has been reported to cause **Wood Rot** (which see).

XYLOPOROSIS. A virus disease causing pitting of the woody cylinder and decline of the top in certain stock and scion combinations. *SYN.*—cachexia.

In 1955, a sampling of trees indicated that 7 out of every 10 trees in Florida carried the virus of xyloporosis (61). Yet the economic consequences have not been serious (except in the growing of Orlando tangelos), because most trees in the state are on tolerant rootstocks and have tops of tolerant scion varieties. Xyloporosis virus is destructive only to susceptible varieties used as stocks or scions. With the use of xyloporosis-free budwood, as made available by the Florida Citrus Budwood Registration Program, even susceptible varieties may now be grown with impunity.

Plants affected

The following varieties show marked susceptibility to xyloporosis: most tangelos (Orlando is particularly susceptible) (55); some tangors (252); various other mandarin hybrids (55, 252); certain types of mandarin, e.g., Satsuma (252); Murcott

(199, 253); *Citrus macrophylla* (51); Leonardy grapefruit (actually a tangelo) (252); Palestine, Columbian, Butwal, and Florida varieties of sweet lime (58, 264); and lime hybrids, e.g., Rangpur lime (59). Some rough-lemon hybrids when used as rootstocks exhibit similar pitting and tree decline symptoms but only after many years (85, 181, 193).

Symptoms

The definitive symptom of xyloporosis is a particular type of pitting in the woody cylinder. The pitting is seen by removing a flap of bark from across the bud union. Pits in the face of the woody cylinder are chisel- or cone-shaped and measure up to ¼ in. across (Fig. 85). Corresponding pegs occur on the inner face of the bark. Pitting is most prominent at the bud union and diminishes with distance from the union. Symptoms are visible as early as 18 months from budding.

A brown, gummy staining accompanies pitting in certain varieties, e.g., Orlando tangelo (Fig. 85B) but is not conspicuously present in others, e.g., Palestine sweet lemon (Fig. 85A). Occasionally, there is a small amount of inverse pitting near the bud union, resembling that produced by tristeza.

Pitting may be produced by other agencies. Not to be mistaken for xyloporosis pitting are: the elongate channels in the woody cylinder caused by tristeza virus in grapefruit and Key lime; the several pits arranged in clusters that are produced by aborted adventitious shoots and roots; and the pits that underlie scale insects on the trunks of young trees (85, 137).

The extent to which the wood is pitted is a measure of the susceptibility of the variety, but the extent to which the top declines is a function of the amount of necrosis produced in phloem tissues. As a result of vascular derangement, the tree becomes stunted, the top declines, and the roots decay. The trunks of affected trees tend to incline because of an elasticity of the woody cylinder. The bark in the vicinity of the bud union may in time crack, scale, and rot (Fig. 86). In the case of susceptible stocks, the vascular necrosis at the bud union prevents the downward passage of food; in time, the impounded food causes an overgrowth of the trunk above the union.

Cause

The cause of xyloporosis is a virus that enters the tree by way of infected buds used during propagation. Studies to date have failed to demonstrate virus transmission through seed (60) or by vectors (239).

The existence of xyloporosis virus can be determined by inserting buds from the suspect tree into indicator plants, customarily Orlando tangelo or Palestine sweet lime. The presence of the virus is established when flecks of gum develop on the inner face of the bark at the bud union. This symptom appears 6–36 months after budding. Pitting symptoms develop later (59).

Control

Affected trees cannot be cured and should be removed as soon as maintenance costs exceed returns. The disease may be avoided by propagating with xyloporosis-free budwood or budwood from seedling trees. In the absence of xyloporosis-free budwood, tolerant scion varieties may be used provided they are budded on rootstock varieties that are also tolerant.

FIG. 85. *XYLOPOROSIS.* A. Pitting of the woody cylinder below the bud union in a 9-year-old sweet-orange tree on Palestine sweet lime. Note absence of conspicuous gumming in this stock variety. B. Pitting accompanied by gumming in the Orlando tangelo interstock portion of a 5-year-old sweet-orange tree on Cleopatra mandarin rootstock. The appearance of the canopy and trunk of this tree is shown in Figure 86.

YELLOW ASCHERSONIA FUNGUS. *Aschersonia goldiana* Sacc. & Ell., one of the so-called "friendly fungi." See **Entomogenous Fungi.**

YELLOW SPOT. A large, bright yellow, interveinal spotting of leaves, with affected areas arranged in a more or less ladder-like progression up each side of the midrib. *SYN.*—molybdenum deficiency.

From the time of its original description in 1908 (110) and until its true nature was discovered in 1951 (308), yellow spot was suspected of being an infectious disease. Now that its cause is known and proper recommendations for its control are widely followed, yellow spot has become more of a curiosity than a problem.

The disorder appeared sporadically throughout

FIG. 86. *XYLOPOROSIS.* Exterior appearance of tree in Figure 85B, showing decline of the top and decay of the bark near the groundline.

the state with trees on the Ridge being most frequently affected. Varieties particularly prone to yellow spot were grapefruit, Temple, King, sweet orange, and tangelo—especially when on grapefruit rootstock.

Symptoms occur in young leaves of the summer flush and consist of small, water-soaked, interveinally disposed areas arranged in a parallel series ascending the leaf blade on both sides of the midrib (Fig. 87). As leaves mature, spots enlarge to an average diameter of ½ in. On upper leaf surfaces, spots are markedly yellow with margins merging into normal green; ultimately, they become brownish. On undersurfaces, the discoloration ranges from an initial oily greenish-brown to an eventual russet-brown. Upper surfaces of lesions are smooth; lower ones, slightly swollen and suffused with gum. In severe cases yellow spot leads to defoliation.

Yellow spot is controlled by spraying affected trees in spring or summer with 1 oz. of either sodium molybdate or ammonium molybdate/100 gal. of water, applied at the rate of 10 gal./tree (309). Regreening of affected areas in young leaves takes place within 5 weeks of application. Older affected leaves sprayed after October do not regreen, but late fall or spring dormant applications will prevent the recurrence of yellow spot in the following flush. Molybdenum deficiency is more likely to occur in soils having a low pH than in soils that are limed to a pH of 5.6 or above. Soil applications of molybdenum have not proved effective.

Since excess molybdenum on citrus peel fed to cattle is toxic, it is recommended that corrective sprays be applied to trees only when symptoms of yellow spot appear.

YELLOW TIPPING. See **Perchlorate Toxicity.**

YELLOW VEIN. A seasonal nitrogen deficiency pattern in leaves. *SYN.*—winter chlorosis.

Midribs and the larger lateral veins of leaves turn yellow for various reasons, including infections and injuries of the bark and roots and the lack of sufficient nitrogen in the soil. Yellow vein is a specific type of veinal chlorosis that at times appears in otherwise healthy trees following periods of particularly cold weather in winter. The amount of yellowing varies from leaf to leaf and from tree to tree. Trees of all ages and varieties may be affected, but yellow vein is found most frequently in young Orlando tangelo and grapefruit trees. In mild cases, only the midribs are chlorotic; in severe cases, entire leaves turn yellow and drop.

Affected leaves are low in nitrogen. When not caused by bark or root damage, the pattern is thought to result from heavy leaching of nitrogen in summer followed, after the fall fertilization, by insufficient rain or irrigation to move nitrogen to the root zone. In such cases, yellow vein can be corrected and defoliation prevented by a liberal application of nitrogen followed by enough rainfall or irrigation to make the nitrogen available to the feeder roots (216). In mild cases, the trouble is self-correcting after the return of warm weather and rains.

YOUNG-TREE DECLINE. A wilt, chlorosis, and die

back commencing as early as 5 years after the planting of citrus trees. *SYN.*—flatwoods decline, blight-like disease, greening-like disease, possibly also **Sandhill Decline** (which see).

In 1965, an apparently new type of decline appeared in young groves of the Ft. Pierce area. In some 5–10-year-old plantings, more than 65 per cent of trees were affected. The disease is now seen to some extent in nearly all recent plantings on the east coast.

Symptoms, varietal susceptibilities, and probable causes of young-tree decline are still matters of debate. Some investigators consider this disease to be an early manifestation of **Blight** (which see). Others point out that, in contrast, blight typically affects older trees, attacks most stock-scion combinations, and causes few if any foliar deficiency patterns. Observers more or less concur, however, that young-tree decline exhibits the following characteristics.

FIG. 87. *YELLOW SPOT.* Large interveinal spotting in grapefruit leaves, indicative of molybdenum deficiency.

Plants affected

Young-tree decline is predominantly a disease of sweet orange trees on rough lemon root. It is noteworthy that symptoms have not been observed to date in lemon trees on rough lemon or in sweet orange trees on sour orange even when growing adjacent to declining sweet orange trees on rough lemon.

Symptoms

The initial characteristic that distinguishes young-tree decline from other declines is the age at which trees begin to show symptoms, viz., 5–10 years after planting. Owing to the newness of this disease, knowledge is lacking on whether older trees may also succumb. Trees are usually rendered unproductive within a year from first appearance of symptoms.

Earliest symptoms are dulling of the foliage, wilting of the leaves, or delayed flushing of the tree. Any one of these effects may occur either on a single branch or throughout the tree. These initial symptoms are usually followed by the appearance in some leaves of a chlorosis resembling zinc deficiency but differing in that yellowed areas are speckled with green dots the size of pinheads to nailheads. Trees with speckled zinc deficiency patterns decline rapidly whereas adjacent trees with typical zinc deficiency patterns (i.e., without green dots unless caused by insect punctures or melanose pustules) decline only to the limited extent usually associated with zinc deficiency. Successive flushes on affected branches produce leaves that are erect, dwarfed, leathery, strap-shaped, and chlorotic. Yellowed leaves are conspicuous during autumn, winter, and spring but disappear in summer, presumably due to shedding. A small percentage of mature fruits are reduced to the size of golf balls and contain curved columellas and aborted seed. As defoliation progresses, twigs begin to die back. Trees do not die completely but sooner or later are removed because of unproductiveness. As in early stages of blight, roots seem at first to be unaffected, though as canopies become thinner, roots commence to starve and deteriorate. The various symptoms associated with young-tree decline are much the same as those of sandhill decline except for the earlier appearance of symptoms in young-tree decline and the occurrence of the two diseases on different soil types.

Cause

The cause of young-tree decline remains unknown. Various possibilities have been explored but so far none of them appears to be the primary cause.

Water damage.—Though water damage has been suspected, roots fail to show symptoms usually associated with excess water.

Soil pH values.—The disease appears not to be related to soil pH. It has been found in soils that range from moderately acid to alkaline. Values obtained from superficial layers, however, may not be as important as those from greater depths, especially under conditions where soils are greatly stratified.

Soil toxicities.—To date, investigations have shown no direct relation between young-tree decline and toxic concentrations of aluminum, iron, sulfides, sodium, and chloride.

Nutritional deficiencies.—Supplemental fertilization with major and minor elements has not cured or halted progression of the disease (68) though in one 5-year fertilizer trial (47), twice as many trees declined when rates of application were one-half of currently recommended levels as when rates were doubled and tripled. Leaves from affected and normal trees have been reported to show no marked differences in content of zinc, manganese, iron, copper, and magnesium (68), but other reports indicate that declining trees, while low in zinc, manganese, and potassium, are high in calcium and sodium (3). Generally, leaf patterns indicate deficiencies of zinc and manganese in the soil, but in the case of young-tree decline they may reflect an inability of diseased feeder roots to absorb elements even when available.

Infectious agents.—Many of the symptoms of young-tree decline are those of greening, a mycoplasma disease not yet known in Florida (194). To determine whether young-tree decline is graft transmissible, numerous infectivity tests have been carried out (68, 173). To date, however, there is no evidence of a bud-transmissible agent.

Physiological factors.—Chromatographic anal-

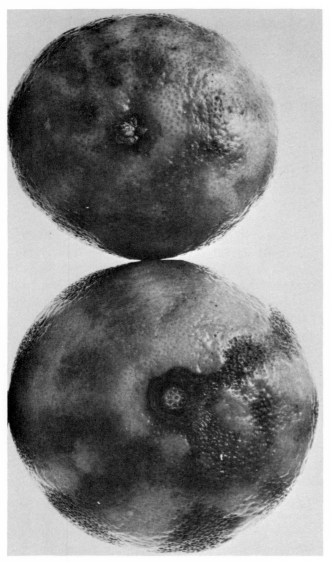

FIG. 88. *ZEBRA-SKIN.* Tangerines affected by a dark-striped rind breakdown. (Photograph by W. Grierson.)

yses of tissues from affected trees have demonstrated the absence of gentisoyl glucose, a substance indicative of infection by the citrus greening mycoplasma (101).

Control

In the absence of information on the cause of young-tree decline, little can be said regarding control except that for the achievement of maximum production severely affected trees should be replaced by new trees. It is not yet known, however, whether resets will thrive or whether they, too, will eventually succumb.

ZEBRA-SKIN. A postharvest rind collapse of tangerines.

The degreening of tangerines may emphasize peel injuries incurred during picking and may also sensitize fruit to certain postethylene handling practices. One type of rind injury, known as zebra-skin (Fig. 88), appears as a dark-colored striping or blotching that coincides with the underlying segments. Affected fruits develop objectionable flavors and are prone to early decay. Symptoms appear several days after fruit has been packed.

Zebra-skin is believed to result from the ethylening of fruit damaged during harvest periods immediately following drought-terminating rains and from the rough handling of fruit after degreening (138, 139).

LITERATURE CITED

1. Adam, D. B., J. McNeil, B. M. Hanson-Merz, D. F. McCarthy, and J. Stokes. 1949. The estimation of latent infection in oranges. Australian J. Sci. Res., B, 2:1–18.
2. Allen, R. 1970. Mechanical transmission of exocortis virus. Calif. Citrograph 55:145–48, 152–53.
3. Anderson, C. A., and D. V. Calvert. 1970. Mineral composition of leaves from citrus trees affected with declines of unknown etiology. Proc. Florida State Hort. Soc. 83:41–45.
4. Anonymous. (Revised annually). Florida citrus spray and dust schedule. Lakeland: State of Florida Department of Citrus.
5. Anonymous. 1960. Index of Plant Diseases in the United States. U.S. Dept. Agr., Agr. Handbook 165.
6. Bach, W. J., and F. A. Wolf. 1928. The isolation of the fungus that causes citrus melanose and the pathological anatomy of the host. J. Agr. Res. 37:243–52.
7. Baines, R. C. 1950. Citrus-root nematode investigations. Calif. Citrograph 35:344–45.
8. Baines, R. C., W. P. Bitters, and O. F. Clarke. 1960. Susceptibility of some species and varieties of citrus and some other rutaceous plants to the citrus nematode. Plant Disease Reptr. 44:281–85.
9. Baines, R. C., L. J. Klotz, T. A. DeWolfe, and R. H. Small. 1962. Chemicals for control of citrus soil pests. Calif. Citrograph 47:342, 359.
10. Baines, R. C., J. P. Martin, T. A. DeWolfe, S. B. Boswell, and M. J. Garber. 1962. Effect of high doses of D-D on soil organisms and the growth and yield of lemon trees. Phytopathology 52:723.
11. Baines, R. C., T. Miyakawa, J. W. Cameron, and R. H. Small. 1969. Infectivity of two biotypes of the citrus nematode on citrus and on some other hosts. J. Nematology 1:150–59.
12. Baker, R. E. D. 1938. Citrus scab disease on grapefruit in Trinidad. Trop. Agr. 15:77–79.
13. Baker, R. E. D. 1938. Studies in the pathogenicity of tropical fungi. II. The occurrence of latent infections in developing fruits. Ann. Botany (London) 8:919–31.
14. Bar-Joseph, M., and G. Loebenstein. 1970. Rapid diagnosis of the citrus tristeza disease by electron microscopy of partially purified preparations. Phytopathology 60:1510–12.
15. Bar-Joseph, M., G. Loebenstein, and J. Cohen. 1970. Partial purification of viruslike particles associated with the citrus tristeza disease. Phytopathology 60:75–78.
16. Barrett, J. T. 1915. Fruit stain and withertip of citrus. Phytopathology 5:293.
17. Bartholomew, E. T. 1937. Endoxerosis, or internal decline, of lemon fruits. Univ. Calif. Agr. Expt. Sta. Bull. 605:1–42.
18. Bartholomew, E. T., W. B. Sinclair, and F. M. Turrell. 1941. Granulation of Valencia oranges. Univ. Calif. Agr. Expt. Sta. Bull. 647.
19. Basson, W. J., and R. E. Schwarz. 1964. Indexing for exocortis and xyloporosis in mother trees of a collection of citrus cultivars. South African J. Agr. Sci. 7:627–32.
20. Batchelor, L. D., and E. C. Calavan. 1952. Lemon strain selection and longevity of the trees. Calif. Citrograph 37:94, 110, 112.
21. Bitancourt, A. A. 1940. A mancha de acaro ou falsa ferrugem das laranjas. O Biologico 6:189–92.
22. Bitancourt, A. A., and A. E. Jenkins. 1936. *Elsinoe fawcetti*, the perfect stage of the citrus scab fungus. Phytopathology 26:393–96.
23. Bitters, W. P. 1972. Reaction of some new citrus hybrids and citrus introductions as rootstocks to inoculations with tristeza virus in California, pp. 112–20. *In* W. C. Price (ed.), Proc. 5th Conf. Intern. Organization Citrus Virol. Gainesville: University of Florida Press.
24. Bitters, W. P., T. Murashige, T. S. Rangan, and E. Nauer. 1972. Investigations on establishing virus-free citrus plants through tissue culture, pp. 267–71. *In* W. C. Price (ed.), Proc. 5th Conf. Intern. Organization Citrus Virol. Gainesville: University of Florida Press.
25. Bitters, W. P., R. G. Platt, J. A. Brusca, and N. W. Dukeshire. 1957. High and low budding of citrus. Calif. Agr. 11(11):5–7, 15.
26. Blazquez, C. H., A. G. Naylor, and D. Hastings. 1966. Sphaeropsis knot of lime. Proc. Florida State Hort. Soc. 79:344–50.
27. Bliss, D. E., and H. S. Fawcett. 1944. The morphology

and taxonomy of *Alternaria citri*. Mycologia 36:469–502.

28. Bridges, G. D. 1966. Tristeza—a growing problem in commercial groves. Citrus Ind. 47(11):33–34.

29. Bridges, G. D., and C. O. Youtsey. 1966. Improved disease control through hot water treatment of citrus seed. Proc. Florida State Hort. Soc. 79:114–15.

30. Bridges, G. D., C. O. Youtsey, and R. R. Nixon. 1965. Observations indicating psorosis transmission by seed of Carrizo citrange. Proc. Florida State Hort. Soc. 78:48–50.

31. Brooks, C. 1944. Stem-end rot of oranges and factors affecting its control. J. Agr. Res. 68:363–81.

32. Brooks, T. L., and V. G. Perry. 1967. Pathogenicity of *Pratylenchus brachyurus* to citrus. Plant Disease Reptr. 51:569–73.

33. Brown, G. E. 1968. Experimental fungicides applied preharvest for control of postharvest decay in Florida citrus fruit. Plant Disease Reptr. 52:844–47.

34. Brown, G. E., and L. G. Albrigo. 1970. Grove application of Benlate for control of postharvest citrus decay. Proc. Florida State Hort. Soc. 83:222–25.

35. Brown, G. E., and W. C. Wilson. 1968. Mode of entry of *Diplodia natalensis* and *Phomopsis citri* into Florida oranges. Phytopathology 58:736–39.

36. Calavan, E. C. 1947. Shell bark of lemons. Calif. Citrograph 32:232–33, 263–65.

37. Calavan, E. C. 1957. Wood pocket disease of lemons and seedless limes. Calif. Citrograph 42:265–68.

38. Calavan, E. C. 1961. Ferment gum disease (Rio Grande gummosis) of grapefruit. Calif. Citrograph 46:231–32.

39. Calavan, E. C. 1968. Exocortis, pp. 28–34. *In* J. F. L. Childs (ed.), Indexing Procedures for 15 Virus Diseases of Citrus Trees. U.S. Dept. Agr., Agr. Handbook 333.

40. Calavan, E. C., R. M. Burns, C. J. Barrett, D. W. Christiansen, and R. L. Blue. 1968. Tristeza in lemon on *Citrus macrophylla* rootstock. Calif. Citrograph 53:108, 119, 122.

41. Calavan, E. C., D. W. Christiansen, and C. N. Roistacher. 1963. Symptoms associated with tatter leaf virus infection of Troyer citrange. Plant Disease Reptr. 47:971–75.

42. Calavan, E. C., R. M. Pratt, B. W. Lee, and J. P. Hill. 1968. Tristeza related to decline of orange trees on citrange rootstock. Calif. Citrograph 53:75, 84–88, 90.

43. Calavan, E. C., and L. G. Weathers. 1954. Fungi and shell bark of lemon. Calif. Agr. 8(6):10–11.

44. Calavan, E. C., and L. G. Weathers. 1959. The distribution of exocortis virus in California citrus, pp. 151–53. *In* J. M. Wallace (ed.), Citrus Virus Diseases. Berkeley: University of California Division of Agricultural Sciences.

45. Calavan, E. C., and L. G. Weathers. 1959. Transmission of a growth-retarding factor in Eureka lemon trees, pp. 167–77. *In* J. M. Wallace (ed.), Citrus Virus

Diseases. Berkeley: University of California Division of Agricultural Sciences.

46. Calavan, E. C., L. G. Weathers, and D. W. Christiansen. 1968. Effect of exocortis on production and growth of Valencia orange trees on trifoliate orange rootstock, pp. 101–4. *In* J. F. L. Childs (ed.), Proc. 4th Conf. Intern. Organization Citrus Virol. Gainesville: University of Florida Press.

47. Calvert, D. V. 1969. Effects of rate and frequency of fertilizer applications on growth, yield and quality factors of young 'Valencia' orange trees. Proc. Florida State Hort. Soc. 82:1–7.

48. Calvert, D. V., and H. J. Reitz. 1965. Salinity of water for sprinkler irrigation of citrus. Proc. Florida State Hort. Soc. 78:73–78.

49. Cameron, J. W., R. C. Baines, and O. F. Clarke. 1954. Resistance of hybrid seedlings of the trifoliate orange to infestation by the citrus nematode. Phytopathology 44:456–58.

50. Carpenter, J. B., and J. R. Furr. 1962. Evaluation of tolerance to root rot caused by *Phytophthora parasitica* in seedlings of *Citrus* and related genera. Phytopathology 52:1277–85.

51. Carpenter, J. B., and J. R. Furr. 1967. Susceptibility of *Citrus macrophylla* 'Alemow' to cachexia virus. Plant Disease Reptr. 51:525–27.

52. Chapot, H. 1963. Le pourridie a *Clitocybe* des agrumes. Al Awamia (Morocco) No. 9:79–87.

53. Childs, J. F. L. 1950. Organic vs. copper fungicides for control of melanose. Phytopathology 40:719–25.

54. Childs, J. F. L. 1950. Rio Grande gummosis. Its occurrence in Florida citrus. Proc. Florida State Hort. Soc. 63:32–36.

55. Childs, J. F. L. 1952. Cachexia, a bud-transmitted disease and the manifestation of phloem symptoms in certain varieties of citrus, citrus relatives and hybrids. Proc. Florida State Hort. Soc. 64:47–51.

56. Childs, J. F. L. 1953. Concentric canker and wood rot of citrus associated with *Fomes applanatus* in Florida. Phytopathology 43:99–100.

57. Childs, J. F. L. 1953. Observations on citrus blight. Proc. Florida State Hort. Soc. 66:33–37.

58. Childs, J. F. L. 1959. Xyloporosis and cachexia—their status as citrus virus diseases, pp. 119–24. *In* J. M. Wallace (ed.), Citrus Virus Diseases. Berkeley: University of California Division of Agricultural Sciences.

59. Childs, J. F. L. 1968. Cachexia (xyloporosis), pp. 16–19. *In* J. F. L. Childs (ed.), Indexing Procedures for 15 Virus Diseases of Citrus Trees. U.S. Dept. Agr., Agr. Handbook 333.

60. Childs, J. F. L. 1968. A review of the cachexia-xyloporosis situation, pp. 83–88. *In* J. F. L. Childs (ed.), Proc. 4th Conf. Intern. Organization Citrus Virol. Gainesville: University of Florida Press.

61. Childs, J. F. L., G. R. Grimm, T. J. Grant, L. C. Knorr, and G. Norman. 1955. The incidence of xyloporosis

(cachexia) in certain Florida citrus varieties. Proc. Florida State Hort. Soc. 68:77–82.

62. Childs, J. F. L., and R. E. Johnson. 1966. Preliminary report of seed transmission of psorosis virus. Plant Disease Reptr. 50:81–83.

63. Childs, J. F. L., and R. C. J. Koo. 1970. Rio Grande gummosis, a physiological disease of citrus trees. (Unpublished manuscript.)

64. Childs, J. F. L., L. E. Kopp, and R. E. Johnson. 1965. A species of *Physoderma* present in *Citrus* and related species. Phytopathology 55:681–87.

65. Childs, J. F. L., G. G. Norman, and J. L. Eichhorn. 1958. A color test for exocortis infection in *Poncirus trifoliata*. Phytopathology 48:426–32.

66. Clausen, R. E. 1912. A new fungus concerned in wither tip of varieties in *Citrus medica*. Phytopathology 2:217–34.

67. Cobb, N. A. 1918. Estimating the nema population of soil. U.S. Dept. Agr., Agr. Tech. Circ. 1.

68. Cohen, M. 1968. Citrus blight and a "blight-like" disease. Citrus Ind. 49(7):12–13, 16, 26.

69. Cohen, M., and H. C. Burnett. 1961. Tristeza in Florida, pp. 107–12. *In* W. C. Price (ed.), Proc. 2nd Conf. Intern. Organization Citrus Virol. Gainesville: University of Florida Press.

70. Cohen, M., G. R. Grimm, and F. W. Bistline. 1964. Foot rot in young groves. Proc. Florida State Hort. Soc. 77:45–52.

71. Cohen, M., and L. C. Knorr. 1953. Present status of tristeza in Florida. Proc. Florida State Hort. Soc. 66:20–22.

72. Cohen, M., and L. C. Knorr. 1954. Honeycombing— a macroscopic symptom of tristeza in Florida. Phytopathology 44:485.

73. Cohen, M., G. D. Ruehle, and F. B. Lincoln. 1961. Influence of some virus and genetic conditions on the growth of Tahiti lime. Proc. Florida State Hort. Soc. 74:24–29.

74. Cohn, E. 1964. Penetration of the citrus nematode in relation to root development. Nematologica 10:594–600.

75. Cohn, E. 1965. On the feeding and histopathology of the citrus nematode. Nematologica 11:47–54.

76. Cohn, E. 1966. The development of the citrus nematode on some of its hosts. Nematologica 11:593–600.

77. Cohn, E. 1969. The citrus nematode, *Tylenchulus semipenetrans* Cobb, as a pest of citrus in Israel, pp. 1013–17. *In* H. D. Chapman (ed.), Proc. 1st Intern. Citrus Symp., vol. 2. Riverside: University of California.

78. Cohn, E., G. Minz, and S. P. Monselise. 1965. The distribution, ecology and pathogenicity of the citrus nematode in Israel. Israel J. Agr. Res. 15:187–200.

79. Conover, R. A. 1950. Studies of stylar-end rot of Tahiti limes. Proc. Florida State Hort. Soc. 63:236–40.

80. Cotton, A. D. 1914. The genus *Atichia*. Roy. Bot. Gardens, Kew, Bull. Misc. Inform. 2:54–63.

81. Darley, E. F., and W. D. Wilbur. 1954. Some relationships of carbon disulfide and *Trichoderma viride* in the control of *Armillaria mellea*. Phytopathology 44:485.

82. Darnell-Smith, G. P., and E. Mackinnon. 1914. Fungus and other diseases of citrus trees. Agr. Gaz. N. S. Wales 25:945–54.

83. Deszyck, E. J., and J. W. Sites. 1954. The effect of lead arsenate sprays on quality and maturity of Ruby Red grapefruit. Proc. Florida State Hort. Soc. 67:38–42.

84. DuCharme, E. P. 1951. Cancrosis B of lemons. Citrus Mag. 13(9):18–20.

85. DuCharme, E. P., and L. C. Knorr. 1954. Vascular pits and pegs associated with diseases in citrus. Plant Disease Reptr. 38:127–42.

86. Eaks, I. L. 1955. The physiological breakdown of the rind of lime fruits after harvest. Proc. Am. Soc. Hort. Sci. 66:141–45.

87. Eaks, I. L. 1960. Physiological studies of chilling injury in citrus fruits. Plant Physiol. 35:632–36.

88. Erickson, L. C. 1968. The general physiology of citrus, pp. 86–126. *In* W. Reuther, L. D. Batchelor, and H. J. Webber (eds.), The Citrus Industry, vol. 2. Berkeley: University of California Press.

89. Fawcett, H. S. 1907. Scaly bark, citrus scab, gumming of citrus, fungi parasitic on citrus whitefly. Florida Agr. Expt. Sta. Rept. for Fiscal Year ending June 30, 1907, pp. 43–49.

90. Fawcett, H. S. 1911. Scaly bark or nail-head rust of citrus. Florida Agr. Expt. Sta. Bull. 106.

91. Fawcett, H. S. 1913. Report of former plant pathologist. Florida Agr. Expt. Sta. Ann. Rept. (1912), pp. 65–92.

92. Fawcett, H. S. 1924. Shell bark (decorticosis) of lemon trees; some investigations and observations. Calif. Citrograph 9:330.

93. Fawcett, H. S. 1936. Citrus Diseases and Their Control. N.Y.: McGraw-Hill.

94. Fawcett, H. S., and A. A. Bitancourt. 1940. Occurrence, pathogenicity and temperature relations of *Phytophthora* species on *Citrus* in Brazil and other South American countries. Arquiv. Inst. Biol. 11:107–18.

95. Fawcett, H. S., and A. A. Bitancourt, 1943. Comparative symptomatology of psorosis varieties on citrus in California. Phytopathology 33:837–64.

96. Fawcett, H. S., and L. J. Klotz. 1948. Diseases and their control, pp. 495–596. *In* L. D. Batchelor and H. J. Webber (eds.), The Citrus Industry, vol. 2. Berkeley: University of California Press.

97. Feder, W. A. 1968. Differential susceptibility of selections of *Poncirus trifoliata* to attack by the citrus nematode, *Tylenchulus semipenetrans*. Israel J. Agr. Res. 18:175–79.

98. Feder, W. A., and P. C. Hitchins. 1966. Twig gumming

and dieback of the 'Robinson' tangerine. Plant Disease Reptr. 50:429–30.

99. Feldman, A. W., G. D. Bridges, R. W. Hanks, and H. C. Burnett. 1971. Effectiveness of the chromatographic method for detecting exocortis virus infection in *Poncirus trifoliata*. Phytopathology 61:1338–41.

100. Feldman, A. W., and R. W. Hanks. 1968. Identification and quantification of phenolics in the leaves and roots of healthy and exocortis-infected citrus, pp. 292–98. *In* J. F. L. Childs (ed.), Proc. 4th Conf. Intern. Organization Citrus Virol. Gainesville: University of Florida Press.

101. Feldman, A. W., and R. W. Hanks. 1969. The occurrence of a gentisic glucoside in the bark and albedo of virus-infected citrus trees. Phytopathology 59:603–6.

102. Feldman, A. W., R. W. Hanks, and S. M. Garnsey. 1972. Localization and detection of coumarins in exocortis-virus-infected citron, pp. 239–44. *In* W. C. Price (ed.), Proc. 5th Conf. Intern. Organization Citrus Virol. Gainesville: University of Florida Press.

103. Feldmesser, J., J. F. L. Childs, and R. V. Rebois. 1964. Occurrence of plant-parasitic nematodes in citrus blight areas. Plant Disease Reptr. 48:95–98.

104. Fisher, F. E. 1957. Control of anthracnose on rough lemon seedlings. Plant Disease Reptr. 41:77–78.

105. Fisher, F. E. 1957. Re-evaluation of the etiology of citrus greasy spot. Phytopathology 47:520–21.

106. Fisher, F. E. 1961. Greasy spot and tar spot of citrus in Florida. Phytopathology 51:297–303.

107. Fisher, F. E. 1966. Tar spot of citrus and its chemical control in Florida. Plant Disease Reptr. 50:357–59.

108. Fisher, F. E. 1967. Cladosporium leaf spot of citrus in Florida. Plant Disease Reptr. 51:1070.

109. Fisher, F. E. 1969. Chemical control of scab on citrus in Florida. Plant Disease Reptr. 53:19–22.

110. Floyd, B. F. 1908. Leaf spotting of citrus. Florida Agr. Expt. Sta. Ann. Rept., p. xci.

111. Floyd, B. F. 1917. Dieback, or exanthema of citrus trees. Florida Agr. Expt. Sta. Bull. 140.

112. Ford, H. W. 1964. The effect of rootstock, soil type, and soil pH on citrus root growth in soils subject to flooding. Proc. Florida State Hort. Soc. 77:41–45.

113. Ford, H. W. 1967. Rootstocks for spreading decline areas. *In* Burrowing Nematode in Citrus. Florida Dept. Agr., Div. Plant Ind. Bull. 7.

114. Ford, H. W. 1969. Water management of wetland citrus in Florida, pp. 1759–70. *In* Proc. 1st Intern. Citrus Symp., vol. 3. Riverside: University of California.

115. Fraser, L. 1949. A gummosis disease of citrus in relation to its environment. Proc. Linnean Soc. N. S. Wales 74:v–xviii.

116. Fraser, L. R. 1957. The relation of seedling yellows to tristeza, pp. 57–62. *In* J. M. Wallace (ed.), Citrus Virus Diseases. Berkeley: University of California

Division of Agricultural Sciences.

117. Fraser, L. R., E. C. Levitt, and J. Cox. 1961. Relationship between exocortis and stunting of citrus varieties on *Poncirus trifoliata* rootstock, pp. 34–39. *In* W. C. Price (ed.), Proc. 2nd Conf. Intern. Organization Citrus Virol. Gainesville: University of Florida Press.

118. Fraser, L. R., K. Long, and J. Cox. 1968. Stem pitting of grapefruit—field protection by the use of mild virus strains, pp. 27–31. *In* J. F. L. Childs (ed.), Proc. 4th Conf. Intern. Organization Citrus Virol. Gainesville: University of Florida Press.

119. Frezzi, M. J. 1940. La lepra explosiva del naranjo. Arg. Min. Agr. Bol. Frutas y Hort. 5(46):3–16.

120. Fulton, H. R. 1925. Relative susceptibility of citrus varieties to attack by *Gloeosporium limetticolum* Clausen. J. Agr. Res. 30:629–35.

121. Garnsey, S. M. 1964. Detection of tatter leaf virus of citrus in Florida. Proc. Florida State Hort. Soc. 77:106–9.

122. Garnsey, S. M. 1970. Viruses in Florida's 'Meyer' lemon trees and their effects on other citrus. Proc. Florida State Hort. Soc. 83:66–71.

123. Garnsey, S. M., and M. Cohen. 1965. Response of various citron selections to exocortis infection in Florida. Proc. Florida State Hort. Soc. 78:41–48.

124. Garnsey, S. M., and J. W. Jones. 1967. Mechanical transmission of exocortis virus with contaminated budding tools. Plant Disease Reptr. 51:410–13.

125. Garnsey, S. M., and L. G. Weathers. 1968. Indexing for tatter leaf with cowpea, pp. 80–82. *In* J. F. L. Childs (ed.), Indexing Procedures for 15 Virus Diseases of Citrus Trees. U.S. Dept. Agr., Agr. Handbook 333.

126. Garnsey, S. M., and L. G. Weathers. 1972. Factors affecting mechanical spread of exocortis, pp. 105–11. *In* W. C. Price (ed.), Proc. 5th Conf. Intern. Organization Citrus Virol. Gainesville: University of Florida Press.

127. Garnsey, S. M., and R. Whidden. 1970. Transmission of exocortis virus to various citrus plants by knife-cut inoculation. Phytopathology 60:1292.

128. Gates, C. M., and M. J. Soule, Jr. 1950. A survey of diseases lethal to Tahiti (Persian) limes in Dade County. Proc. Florida State Hort. Soc. 63:225–28.

129. Gaumann, E. 1950. Principles of Plant Infection. N.Y.: Hafner Publishing Co.

130. Gerber, J. F., and J. D. Martsolf. 1966. Protecting citrus from cold damage. Florida Agr. Ext. Serv. Circ. 287.

131. Gibson, I. A. S. 1961. A note on variation between isolates of *Armillaria mellea* (Vahl ex Fr.) Kummer. Trans. Brit. Mycol. Soc. 44:123–28.

132. Gibson, I. A. S., and D. C. M. Corbett. 1964. Variation in isolates from Armillaria root disease in Nyasaland. Phytopathology 54:122–23.

133. Godfrey, G. H. 1945. A gummosis of citrus associated with wood necrosis. Science 102(2640):130.

134. Grant, T. J. 1957. Effect of heat treatment on tristeza and psorosis viruses of citrus. Plant Disease Reptr. 41:232–34.

135. Grant, T. J., and M. K. Corbett. 1961. Mechanical transmission of infectious variegation virus in citrus and non-citrus hosts, pp. 197–204. In W. C. Price (ed.), Proc. 2nd Conf. Intern. Organization Citrus Virol. Gainesville: University of Florida Press.

136. Grant, T. J., and M. K. Corbett. 1964. Properties of citrus variegation virus. Phytopathology 54:946–48.

137. Grant, T. J., G. R. Grimm, and P. Norman. 1959. Symptoms of cachexia in Orlando tangelo, none in sweet lime and false symptoms associated with purple scale infestations. Plant Disease Reptr. 43: 1277–79.

138. Grierson, W., and G. E. Brown. 1966. "Zebra-skin" injury of tangerines as related to pre- and post-harvest handling. Citrus Ind. 47(3):8–10.

139. Grierson, W., and W. F. Newhall. 1960. Degreening of Florida citrus fruits. Florida Agr. Expt. Sta. Bull. 620.

140. Grierson, W., and R. Patrick. 1956. The sloughing disease of grapefruit. Proc. Florida State Hort. Soc. 69:140–42.

141. Griffiths, J. T., and W. L. Thompson. 1957. Insects and mites found on Florida citrus. Florida Agr. Expt. Sta. Bull. 591.

142. Grimm, G. R., and S. M. Garnsey. 1968. Foot rot and tristeza tolerance of Smooth Seville orange from two sources. Proc. Florida State Hort. Soc. 81:84–90.

143. Grimm, G. R., and R. Whidden. 1962. Range of pathogenicity of Florida cultures of the foot rot fungus. Proc. Florida State Hort. Soc. 75:73–74.

144. Haas, A. R. C., and H. J. Quayle. 1935. Copper content of citrus leaves and fruit in relation to exanthema and fumigation injury. Hilgardia 9:143–77.

145. Hannon, C. I. 1958. The citrus nematode. Sunshine State Agr. Res. Rept. (Florida Agr. Expt. Sta.) 3(2):7–8.

146. Hannon, C. I. 1962. The occurrence and distribution of the citrus-root nematode, Tylenchulus semipenetrans Cobb, in Florida. Plant Disease Reptr. 46:451–55.

147. Hannon, C. I. 1965. Longevity of the citrus-root nematode in Florida. Proc. Soil Crop Sci. Soc. Florida 24:158–61.

148. Hara, K. 1962. On the sooty mould of citrus. Trans. Mycol. Soc. Japan 3:104–11.

149. Hart, T. S. 1946. Notes on the identification and growth of certain dodder-laurels. Victorian Natur. 63(1):12–16.

150. Hatton, T. T., Jr., and W. F. Reeder. 1968. Stylar-end breakdown in Persian limes influenced by temperature and bruising. Proc. Florida State Hort. Soc. 81:344–49.

151. Hearn, C. J., and J. F. L. Childs. 1969. A systemic fungicide effective against sour orange scab disease.

152. Hearn, C. J., J. F. L. Childs, and R. Fenton. 1971. Comparison of benomyl and copper sprays for control of sour orange scab of citrus. Plant Disease Reptr. 55:241–43.

153. Hearn, C. J., and R. Fenton. 1970. Benomyl sprays for control of twig dieback of 'Robinson' tangerine. Plant Disease Reptr. 54:869–70.

154. Hedges, F., and L. S. Tenny. 1912. A knot of citrus trees caused by Sphaeropsis tumefaciens. U.S. Dept. Agr. Bur. Plant Ind. Bull. 247.

155. Heinis, J. L. 1962. Leaf spot on Rangpur lime and its control. Plant Disease Reptr. 46:200–201.

156. Hildebrand, E. M. 1947. Stem-end fungi attack immature citrus fruit. Phytopathology 37:433.

157. Hopkins, E. F., and A. A. McCornack. 1964. Decay control. Fla. Agr. Expt. Sta. Bull. 681:25–28.

158. Hume, H. H. 1900. Some citrus troubles. Florida Expt. Sta. Bull. 53:145–73.

159. Jamoussi, B. 1955. Les maladies de deperissement des agrumes. Rev. Mycol. 20, Suppl. colon., 1, pp. 1–47.

160. Jenkins, A. E. 1925. The citrus scab fungus. Phytopathology 15:99–104.

161. Jenkins, A. E. 1931. Development of the citrus-scab organism, Sphaceloma fawcetti. J. Agr. Res. 42:545–58.

162. Jenkins, A. E., L. C. Knorr, and A. A. Bitancourt. 1953. Notes on spot anthracnoses and related subjects. V. Discovery of Tryon's citrus scab in Argentina. Rev. Arg. Agron. 20(4):230–32.

163. Jeppson, L. R., M. J. Jesser, and J. O. Complin. 1955. Control of mites on citrus with chlorobenzilate. J. Econ. Entomol. 48:375–77.

164. Jones, W. W., and T. W. Embleton. 1967. Creasing of orange fruit. Calif. Citrograph 52:398, 408, 410.

165. Jones, W. W., T. W. Embleton, M. J. Garber, and C. B. Cree. 1967. Creasing of orange fruit. Hilgardia 38:231–44.

166. Karsten, G. 1891. Untersuchungen ueber die Familie der Chroolepideen. Ann. Jard. Bot. Buitenzorg 10:1–65.

167. Kay, A. O. 1925. Soil moisture studies in relation to diseased tree conditions in Brevard County. Citrus Ind. 6(8):5–9, 22, 23.

168. Kitajima, E. W., D. M. Silva, A. R. Oliveira, G. W. Muller, and A. S. Costa. 1965. Electron microscopical investigations on tristeza, pp. 1–9. In W. C. Price (ed.), Proc. 3d Conf. Intern. Organization Citrus Virol. Gainesville: University of Florida Press.

169. Klotz, L. J. 1961. Color Handbook of Citrus Diseases. Berkeley: University of California Division of Agricultural Sciences.

170. Klotz, L. J., W. P. Bitters, T. A. DeWolfe, and M. J. Garber. 1967. Orchard tests of citrus rootstocks for resistance to Phytophthora. Calif. Citrograph 53:38, 55.

Plant Disease Reptr. 53:203–5.

171. Klotz, L. J., T. A. DeWolfe, C. N. Roistacher, E. M. Nauer, and J. B. Carpenter. 1960. Heat treatments to destroy fungi in infected seeds and seedlings of citrus. Plant Disease Reptr. 44:858–61.

172. Knorr, L. C. From private correspondence.

173. Knorr, L. C. Unpublished information.

174. Knorr, L. C. 1949. A gall of Tahiti lime and other citrus species caused by dodder. Phytopathology 39:616–20.

175. Knorr, L. C. 1949. Parasitism of citrus in Florida by various species of dodder, including *Cuscuta Boldinghii* Urb., a species newly reported for the United States. Phytopathology 39:411–12.

176. Knorr, L. C. 1950. Algal spot of citrus fruits. Citrus Mag. 12(8):16–17.

177. Knorr, L. C. 1950. Etiological association of a Brevipalpus mite with Florida scaly bark of citrus. Phytopathology 40:15.

178. Knorr, L. C. 1953. Transmission trials with crinkle-scurf of citrus. Plant Disease Reptr. 37:503–7.

179. Knorr, L. C. 1956. Progress of citrus brown rot in Florida, a disease of recent occurrence in the state. Plant Disease Reptr. 40:772–74.

180. Knorr, L. C. 1956. Suscepts, indicators, and filters of tristeza virus, and some differences between tristeza in Argentina and in Florida. Phytopathology 46:557–60.

181. Knorr, L. C. 1957. Re-appraising citrus rootstocks, with particular reference to their susceptibility to virus diseases. III. Rough lemon (*Citrus limon* [L.] Burm. f.). Citrus Mag. 20(2):8–10, 21, 24–25.

182. Knorr, L. C. 1959. Fovea—a disease of concern to Murcott growers. Citrus Ind. 40(6):5–7, 16.

183. Knorr, L. C. 1962. Etiology and control of virus diseases of citrus. Ann. Rept. Florida Agr. Expt. Sta., pp. 215–16.

184. Knorr, L. C. 1962. Malpractices in the citrus nursery that predispose grove trees to disease. Proc. Florida State Hort. Soc. 75:42–48.

185. Knorr, L. C. 1963. Cancroid spot of citrus. Phytopathology 53:1415–18.

186. Knorr, L. C. 1963. Dossier on sweet-orange scab. Citrus Ind. 44(9):7, 9, 12, 26.

187. Knorr, L. C. 1963. Rumple—a new disease of lemon fruits. Plant Disease Reptr. 47:335–39.

188. Knorr, L. C. 1964. A suggestion that the Lee tangerine may be hypersensitive to *Cephaleuros virescens*. Plant Disease Reptr. 48:478–79.

189. Knorr, L. C. 1965. Serious diseases of citrus foreign to Florida. Florida Dept. Agr. Div. Plant Ind. Bull. 5.

190. Knorr, L. C. 1965. Zineb contra-indicated as a control for leprosis in citrus. Trop. Agr. 42:175–76.

191. Knorr, L. C. 1966. Diagnosing tristeza. Citrus Ind. 47(7):14–15.

192. Knorr, L. C. 1966. Sour orange popular despite increasing losses from tristeza. Citrus Ind. 47(5):18, 32.

193. Knorr, L. C. 1967. Citrus diseases and the need for validated rootstock seed sources. Citrus Ind. 48(2):18–19.

194. Knorr, L. C. 1967. Greening disease: What is it, and is it present in Florida? Citrus Ind. 48(4):13–14, 18–19, 21.

195. Knorr, L. C. 1968. Podagra, pp. 77–79. *In* J. F. L. Childs (ed.), Indexing Procedures for 15 Virus Diseases of Citrus Trees. U.S. Dept. Agr., Agr. Handbook 333.

196. Knorr, L. C. 1968. Studies on the etiology of leprosis in citrus, pp. 321–41. *In* J. F. L. Childs (ed.), Proc. 4th Conf. Intern. Organization Citrus Virol. Gainesville: University of Florida Press.

197. Knorr, L. C. 1968. Transmission trials with virus-like diseases of citrus in Florida, pp. 325–31. *In* J. F. L. Childs (ed.), Proc. 4th Conf. Intern. Organization Citrus Virol. Gainesville: University of Florida Press.

198. Knorr, L. C., and J. F. L. Childs. 1957. Occurrence of wood pocket (blotch), chimeric breakdown, and endoxerosis in Florida, with particular reference to the Tahiti lime. Proc. Florida State Hort. Soc. 70:75–81.

199. Knorr, L. C., and R. J. Collins. 1966. Murcotts in distress. Citrus Ind. 47(6):17–19.

200. Knorr, L. C., and H. A. Denmark. 1970. Injury to citrus by the mite *Brevipalpus phoenicis*. J. Econ. Entomol. 63:1996–98.

201. Knorr, L. C., H. A. Denmark, and H. C. Burnett. 1968. Occurrence of Brevipalpus mites, leprosis, and false leprosis on citrus in Florida. Florida Entomol. 51:11–17.

202. Knorr, L. C., and E. P. DuCharme. 1953. Thirty-three Florida rough lemon seed sources tested for tolerance to tristeza. Citrus Mag. 15(12):24–25.

203. Knorr, L. C., E. P. DuCharme, and J. N. Busby. 1954. Discovery of exocortis in Florida citrus. Plant Disease Reptr. 38:12–13.

204. Knorr, L. C., and W. Eberling. 1961. World citrus problems. II. The Gaza Strip. F.A.O. Plant Protect. Bull. 9:115–20.

205. Knorr, L. C., and R. C. J. Koo. 1969. Rumple—a serious rind collapse of lemons in Florida and in Mediterranean countries, pp. 1463–72. *In* Proc. 1st Intern. Citrus Symp., vol. 3. Riverside: University of California.

206. Knorr, L. C., and W. C. Price. 1956. Is stem pitting of grapefruit a threat to the Florida grower? Proc. Florida State Hort. Soc. 69:65–68.

207. Knorr, L. C., and W. C. Price. 1958. Diagnosis and rapid determination of tristeza. *In* Florida Agr. Expt. Sta. Ann. Rept. for year ending June 30, 1958, pp. 252–53.

208. Knorr, L. C., and W. C. Price. 1959. Fovea—a disease of the Murcott. Citrus Mag. 22(1):16–19, 26.

209. Knorr, L. C., and H. J. Reitz. 1959. Exocortis in Florida, pp. 141–50. *In* J. M. Wallace (ed.), Citrus Virus Diseases. Berkeley: University of California Division

of Agricultural Sciences.

210. Knorr, L. C., R. F. Suit, and E. P. DuCharme. 1957. Handbook of citrus diseases in Florida. Florida Agr. Expt. Sta. Bull. 587.

211. Knorr, L. C., and W. L. Thompson. 1954. Spraying trials for the control of Florida scaly bark in citrus. Plant Disease Reptr. 38:143–46.

212. Knorr, L. C., B. N. Webster, and G. Malaguti. 1960. Injuries in citrus attributed to Brevipalpus mites, including Brevipalpus gall, a newly reported disorder in sour-orange seedlings. F.A.O. Plant Protect. Bull. 8:141–48.

213. Lawrence, F. P. 1960. Treatment of cold-injured citrus trees. Florida Agr. Ext. Serv. Circ. 174A.

214. Leonard, C. D., and H. B. Graves, Jr. 1966. Effect of air-borne fluorides on 'Valencia' orange yields. Proc. Florida State Hort. Soc. 79:79–86.

215. Leonard, C. D., and H. B. Graves, Jr. 1970. Some effects of air-borne fluorine on growth and yield of six citrus varieties. Proc. Florida State Hort. Soc. 83:34–41.

216. Leonard, C. D., and I. Stewart. 1961. Yellow-vein in citrus. Citrus Ind. 42(2):5–7, 12–13, 15.

217. Lucie-Smith, M. N. (n.d.) The lime disease problem. Trinidad & Tobago Dept. Agr.

218. McClean, A. P. D. 1957. Tristeza virus of citrus: evidence for absence of seed transmission. Plant Disease Reptr. 41:821.

219. McClean, A. P. D., and J. E. van der Plank. 1955. The role of seedling yellows and stem pitting in tristeza of citrus. Phytopathology 45:222–24.

220. McCornack, A. A. 1970. Status of postharvest fungicides for citrus fruit. Proc. Florida State Hort. Soc. 83:229–32.

221. McCornack, A. A., and G. E. Brown. 1969. Benlate, an experimental postharvest citrus fungicide. Proc. Florida State Hort. Soc. 82:235–38.

222. McCornack, A. A., and W. Grierson. 1965. Practical measures for control of rind breakdown of oranges. Florida Agr. Ext. Serv. Circ. 286.

223. McOnie, K. C. 1964. Apparent absence of Guignardia citricarpa Kiely from localities where citrus black spot is absent. South African J. Agr. Sci. 7:347–54.

224. Majorana, G., and G. P. Martelli. 1968. Comparison of citrus infectious variegation and citrus crinkly-leaf virus isolates from Italy and California, pp. 273–80. In J. F. L. Childs (ed.), Proc. 4th Conf. Intern. Organization Citrus Virol. Gainesville: University of Florida Press.

225. Martin, J. T., E. A. Baker, and R. J. W. Byrde. 1966. The fungitoxicities of cuticular and cellular components of citrus lime leaves. The fungitoxicities of plant furocoumarins. Ann Appl. Biol. 57:491–500, 501–8.

226. Martinez, A. L. 1972. Exocortis virus, the cause of bark disorder of Sziwuikom mandarin in the Philippines. In Abstracts of papers for presentation at the 5th Conf. Intern. Organization Citrus Virol. (Mimeo.)

227. Minz, G. 1946. Diplodia natalensis, its occurrence on flowers, button and stem-end of Shamouti orange and its relation to stem-end rot and fruit drop. Palestine J. Bot., Rehovot Ser. 5:152–68.

228. Minz, G., and Y. Ben-Meir. 1952. Comparative pathogenicity of potential causal agent of stem-end rot in Shamouti oranges. Ktavim 2–3:39–40.

229. Moherek, E. A. 1971. Disease control in Florida citrus with Difolatan fungicide. Proc. Florida State Hort. Soc. 83:59–65.

230. Moje, W. 1959. Structure and nematocidal activity of allylic and acetylenic halides. J. Agr. Food Chem. 7:702–7.

231. Moreira, S. 1968. Growing citrus in the presence of tristeza, pp. 41–44. In J. F. L. Childs (ed.), Proc. 4th Conf. Intern. Organization Citrus Virol. Gainesville: University of Florida Press.

232. Muller, G. W., and A. S. Costa. 1968. Further evidence on protective interference in citrus tristeza, pp. 71–82. In J. F. L. Childs (ed.), Proc. 4th Conf. Intern. Organization Citrus Virol. Gainesville: University of Florida Press.

233. Muma, M. H., A. G. Selhime, and H. A. Denmark. 1972. An annotated list of predators and parasites associated with insects and mites on Florida citrus. Florida Agr. Expt. Sta. Tech. Bull. 634 (revision in press).

234. Nadel, M. 1944. Anatomical study of the button of Shamouti oranges in relation to stem-end rot. Palestine J. Bot., Rehovot Ser. 4:166–70.

235. Nolla, J. A. B. 1926. The anthracnose of citrus fruits, mango, and avocado. J. Dept. Agr. Puerto Rico 10(2):25–63.

236. Norman, G. G. 1959. Florida State Plant Board program for virus-free budwood, pp. 237–42. In J. M. Wallace (ed.), Citrus Virus Diseases. Berkeley: University of California Division of Agricultural Sciences.

237. Norman, G. G. 1965. The incidence of exocortis virus in Florida citrus varieties, pp. 124–27. In W. C. Price (ed.), Proc. 3d Conf. Intern. Organization Citrus Virol. Gainesville: University of Florida Press.

238. Norman, G. G., W. C. Price, T. J. Grant, and H. Burnett. 1961. Ten years of tristeza in Florida. Proc. Florida State Hort. Soc. 74:107–11.

239. Norman, P. A., and J. F. L. Childs. 1963. Attempted transmission of xyloporosis of citrus with insects. Proc. Florida State Hort. Soc. 76:48–50.

240. Norman, P. A., and T. J. Grant. 1956. Transmission of tristeza virus by aphids in Florida. Proc. Florida State Hort. Soc. 69:38–42.

241. O'Bannon, J. H. 1968. The influence of an organic soil amendment on infectivity and reproduction of Tylenchulus semipenetrans on two citrus rootstocks. Phytopathology 58:597–601.

242. O'Bannon, J. H., and F. W. Bistline. 1969. A simple

device for injecting methyl bromide into a replant site. Plant Disease Reptr. 53:799–802.

243. O'Bannon, J. H., H. W. Reynolds, and C. R. Leathers. 1967. Effects of temperature on penetration, development, and reproduction of *Tylenchulus semipenetrans*. Nematologica 12:483–87.

244. O'Bannon, J. H., and A. C. Tarjan. 1969. Increasing yield of Florida citrus through chemical control of the citrus nematode, *Tylenchulus semipenetrans*, pp. 991–98. *In* Proc. 1st Intern. Citrus Symp., vol. 2. Riverside: University of California.

245. O'Bannon, J. H., and A. L. Taylor. 1967. Control of nematodes on citrus seedlings by chemical bare-root dip. Plant Disease Reptr. 51:995–98.

246. Oberbacher, M. F. 1954. A chlorosis of citrus produced by biuret as an impurity in urea. Proc. Florida State Hort. Soc. 67:67–69.

247. Oberbacher, M. F. 1965. A method to predict the post-harvest incidence of oleocellosis in lemons. Proc. Florida State Hort. Soc. 78:237–40.

248. Oberbacher, M. F., and L. C. Knorr. 1965. Increase of rumple and decay in lemon fruits during storage. Proc. Am. Soc. Hort. Sci. 86:260–66.

249. Oberholzer, P. C. J., I. Mathews, and S. F. Stiemie. 1949. The decline of grapefruit trees in South Africa. A preliminary report on so-called "stem pitting." Union S. Africa Dept. Agr. Sci. Bull. 297.

250. Oliveira, A. R., G. W. Muller, and A. S. Costa. 1972. Serology of citrus tristeza virus. *In* Abstracts of papers for presentation at the 5th Conf. Intern. Organization Citrus Virol. (Mimeo.)

251. Olson, E. O. 1951. Tip blight of citrus seedlings in the Lower Rio Grande Valley. Proc. Rio Grande Valley Hort. Inst. 5:72–76.

252. Olson, E. O. 1952. Investigations of citrus rootstock diseases in Texas. Proc. Rio Grande Valley Hort. Inst. 6:28–34.

253. Olson, E. O. 1960. Xyloporosis (cachexia or fovea) disease of Murcott Honey "orange" in Texas. J. Rio Grande Valley Hort. Soc. 14:26–28.

254. Olson, E. O., A. Shull, and G. Buffington. 1961. Evaluation of indicators for xyloporosis and exocortis in Texas, pp. 159–65. *In* W. C. Price (ed.), Proc. 2nd Conf. Intern. Organization Citrus Virol. Gainesville: University of Florida Press.

255. Pantastico, E. B., J. Soule, and W. Grierson. 1968. Chilling injury in tropical and subtropical fruits. II. Limes and grapefruit. Am. Soc. Hort. Sci., Trop. Reg. Proc. 12:171–83.

256. Peltier, G. L. 1920. Influence of temperature and humidity on the growth of *Pseudomonas citri* and its host plants and on infection and development of the disease. J. Agr. Res. 20:447–506.

257. Poucher, C. Personal communication.

258. Poucher, C., H. W. Ford, R. F. Suit, and E. P. DuCharme. 1967. Burrowing nematode in citrus. Florida Dept. Agr. Bull. 7.

259. Pratt, R. M. 1958. Florida Guide to Citrus Insects, Diseases and Nutritional Disorders in Color. Gainesville: Florida Agricultural Experiment Station.

260. Price, W. C. 1965. Transmission of psorosis virus by dodder, pp. 162–66. *In* W. C. Price (ed.), Proc. 3d Conf. Intern. Organization Citrus Virol. Gainesville: University of Florida Press.

261. Price, W. C. 1968. Translocation of tristeza and psorosis viruses, pp. 52–58. *In* J. F. L. Childs (ed.), Proc. 4th Conf. Intern. Organization Citrus Virol. Gainesville: University of Florida Press.

262. Pujol, A. R. 1966. Transmission de psorosis a traves de la semilla de citrange Troyer. Estac. Expt. Agropecuaria de Concordia, INTA Serie Technica No. 10.

263. Reese, R. L. 1967. Improved fertilizer material gives reduced perchlorate toxicity symptoms. Proc. Florida State Hort. Soc. 80:15–19.

264. Reichert, I., and J. Perlberger. 1934. Xyloporosis, the new citrus disease. Jewish Agency Palestine, Agr. Res. Sta., Rehovot, Bull. 12:1–50.

265. Reitz, H. J., and L. C. Knorr. 1968. Effect of crinkle-scurf of citrus on tree growth, productivity, and fruit quality. Plant Disease Reptr. 52:31–33.

266. Reitz, H. J., C. D. Leonard, I. Stewart, R. C. J. Koo, D. V. Calvert, C. A. Anderson, P. F. Smith, and G. K. Rasmussen. 1964. Recommended fertilizers and nutritional sprays for citrus. Florida Agr. Expt. Sta. Bull. 536B.

267. Reuther, W., and P. F. Smith. 1954. Toxic effects of accumulated copper in Florida soils. Proc. Soil Sci. Soc. Florida 14:17–23.

268. Reynolds, H. W., and J. H. O'Bannon. 1963. Decline of grapefruit trees in relation to citrus nematode populations and tree recovery after chemical treatment. Phytopathology 53:1011–15.

269. Rhoads, A. S. 1936. Blight—a non-parasitic disease of citrus trees. Florida Agr. Expt. Sta. Bull. 296.

270. Rhoads, A. S. 1945. A comparative study of two closely related root-rot fungi, *Clitocybe tabescens* and *Armillaria mellea*. Mycologia 37:741–66.

271. Rhoads, A. S. 1950. Clitocybe root rot of woody plants in the southeastern United States. U.S. Dept. Agr. Circ. 853.

272. Rhoads, A. S., and E. F. DeBusk. 1931. Diseases of citrus in Florida. Florida Agr. Expt. Sta. Bull. 229.

273. Roistacher, C. N., E. C. Calavan, and R. L. Blue. 1969. Citrus exocortis virus—chemical inactivation on tools, tolerance to heat and separation of isolates. Plant Disease Reptr. 53:333–36.

274. Roistacher, C. N., and E. M. Nauer. 1964. A comparison of certain sweet orange varieties as indicators for concave gum and psorosis viruses. Plant Disease Reptr. 48:56–59.

275. Rolfs, P. H. (*In* Porcher, E. P.) 1903. Diseases and insects of the citrus. Proc. Florida State Hort. Soc. 16:44–45.

276. Rolfs, P. H., H. S. Fawcett, and B. F. Floyd. 1911.

Diseases of citrus fruits. Florida Agr. Expt. Sta. Bull. 108:27–47.

277. Roth, G. 1966. The pathogenicity of *Diplodia natalensis* Pole-Evans and *Phytophthora citrophthora* (Sm. and Sm.) Leon. to citrus rootstocks. Phytopathol. Z. 57:201–15.

278. Ruehle, G. D. 1937. A strain of *Alternaria citri* Ellis and Pierce causing a leaf spot of rough lemon in Florida. Phytopathology 27:863–65.

279. Ruehle, G. D. 1941. Algal leaf and fruit spot of guava. Phytopathology 31:95–96.

280. Ruehle, G. D., and W. A. Kuntz. 1933. Melanose and stem-end rot of citrus. Florida Agr. Expt. Sta. Ann. Rept. 1933, pp. 139–40.

281. Ruehle, G. D., and W. A. Kuntz. 1940. Melanose of citrus and its commercial control. Florida Agr. Expt. Sta. Bull. 349.

282. Ruehle, G. D., and W. L. Thompson. 1939. Commercial control of citrus scab in Florida. Florida Agr. Expt. Sta. Bull. 337.

283. Salama, S. B., W. Grierson, and M. F. Oberbacher. 1965. Storage trials with limes, avocados, and lemons in modified atmospheres. Proc. Florida State Hort. Soc. 78:353–58.

284. Salerno, M. 1963. Il "raggrinzimento della buccia:" grave alterazione dei frutti di limone. Tec. Agricola 15:507–11.

285. Salerno, M. 1965. Osservazione sulla eziologia del "raggrinzimento della buccia" dei frutti di limone. Riv. Patol. Veg., Ser. 4, 1:33–40.

286. Salerno, M., G. Perrotta, and M. Benintende. 1968. L'incidenza del 'raggrinzimento della buccia' in rapporto ad alcuni livelli nutritivi in piante di limone. Riv. Patol. Veg., Pavia, Ser. 4, 4(3):201–10.

287. Salibe, A. A., and S. Moreira. 1965. New test varieties for exocortis virus, pp. 119–23. In W. C. Price (ed.), Proc. 3d Conf. Intern. Organization Citrus Virol. Gainesville: University of Florida Press.

288. Salibe, A. A., and S. Moreira. 1965. Tahiti lime bark disease is caused by exocortis virus, pp. 143–47. In W. C. Price (ed.), Proc. 3d Conf. Intern. Organization Citrus Virol. Gainesville: University of Florida Press.

289. Salibe, A. A., S. Moreira, and O. Rodriguez. 1972. Performance of selections of trifoliate orange and trifoliate orange hybrids as rootstocks for citrus in the presence of tristeza virus, pp. 124–27. In W. C. Price (ed.), Proc. 5th Conf. Intern. Organization Citrus Virol. Gainesville: University of Florida Press.

290. Schneider, H. 1954. Anatomy of bark of bud union, trunk, and roots of quick-decline-affected sweet orange trees on sour orange rootstock. Hilgardia 22: 567–81.

291. Schneider, H. 1957. Anatomical response of some citrus species to tristeza virus. Phytopathology 47: 534.

292. Schneider, H., and R. C. Baines. 1964. *Tylenchulus semipenetrans*: parasitism and injury to orange tree roots. Phytopathology 54:1202–6.

293. Schwarz, R. E. 1965. Evaluation of the methods for indexing exocortis virus of citrus. South African J. Agr. Sci. 8:593–98.

294. Schwarz, R. E. 1968. Indexing of greening and exocortis through fluorescent marker substances, pp. 118–24. In J. F. L. Childs (ed.), Proc. 4th Conf. Intern. Organization Citrus Virol. Gainesville: University of Florida Press.

295. Semancik, J. S., and L. G. Weathers. 1970. Properties of the infectious forms of exocortis virus of citrus. Phytopathology 60:732–36.

296. Simanton, W. A., and L. C. Knorr. 1969. Aphid populations in relationship to tristeza in Florida citrus. Florida Entomol. 52:21–27.

297. Sinclair, W. B., and V. A. Jolliffe. 1961. Chemical changes in the juice vesicles of granulated Valencia oranges. J. Food Sci. 26:276–82.

298. Singer, R. 1949. The "Agaricales" (mushrooms) in modern taxonomy. Lilloa (Univ. Nac. Tucuman, Argentina).

299. Sites, J. W., and E. J. Deszyck. 1952. Effect of varying amounts of potash on yield and quality of Valencia and Hamlin oranges. Proc. Florida State Hort. Soc. 65:92–98.

300. Smith, C. O. 1917. Sour rot of lemon in California. Phytopathology 7:37–41.

301. Smoot, J. J., L. G. Houck, and H. B. Johnson. 1971. Market Diseases of Citrus and Other Subtropical Fruits. U.S. Dept. Agr., Agr. Handbook 398.

302. Standifer, M. S., and V. G. Perry. 1960. Some effects of sting and stubby root nematodes on grapefruit roots. Phytopathology 50:152–56.

303. Steiner, G. 1942. Plant nematodes the grower should know. Proc. Soil Sci. Soc. Florida 4-B:72–117.

304. Stevens, H. E. 1914. Gummosis. Florida Agr. Expt. Sta. Ann. Rept., pp. lvii–lxxi.

305. Stevens, H. E. 1918. Florida citrus diseases. Florida Agr. Expt. Sta. Bull. 150.

306. Stevens, H. E. 1918. Lightning injury to citrus trees in Florida. Phytopathology 8:283–85.

307. Stewart, I., and C. D. Leonard. 1952. The cause of yellow tipping in citrus leaves. Proc. Florida State Hort. Soc. 65:25–27.

308. Stewart, I., and C. D. Leonard. 1952. Molybdenum deficiency in Florida citrus. Nature 170:714–15.

309. Stewart, I., and C. D. Leonard. 1953. Correction of molybdenum deficiency in Florida citrus. Proc. Am. Soc. Hort. Sci. 62:111–15.

310. Stokes, D. E. 1969. *Andropogon rhizomatus* parasitized by a strain of *Tylenchulus semipenetrans* not parasitic to four citrus rootstocks. Plant Disease Reptr. 53:882–85.

311. Stout, R. G. 1964. Specific gravity as a means of estimating juice yields of freeze damaged Valencia oranges. Florida Agr. Expt. Sta. Circ. S-150.

312. Stroyan, H. L. G. 1961. Identification of aphids living on citrus. F.A.O. Plant Protect. Bull. 9:45–65.

313. Stubbs, L. L. 1963. Tristeza-tolerant strains of sour orange. F.A.O. Plant Protect. Bull. 11(1):8–10.

314. Stubbs, L. L. 1964. Transmission and protective inoculation studies with viruses of the citrus tristeza complex. Australian J. Agr. Res. 15:752–70.

315. Stubbs, L. L. 1968. Apparent elimination of exocortis and yellowing viruses in lemon by heat therapy and shoot-tip propagation, pp. 96–99. In J. F. L. Childs (ed.), Proc. 4th Conf. Intern. Organization Citrus Virol. Gainesville: University of Florida Press.

316. Suit, R. F. 1949. Recent experiments on melanose control with reference to organic fungicides and dormant sprays. Citrus Ind. 30(2):8, 18.

317. Suit, R. F. 1952. Comparison of copper fungicides for melanose control. Citrus Mag. 14(7):24–26, 35.

318. Suit, R. F. 1966. Control of algal disease on citrus. Florida Agr. Expt. Sta. Ann. Rept., pp. 260–61.

319. Suit, R. F. 1970. Disease control in citrus. Univ. Florida Inst. Food & Agr. Sci. Proc. Pest Control Conf. 4:122–28.

320. Suit, R. F., and E. P. DuCharme. 1953. The burrowing nematode and other parasitic nematodes in relation to spreading decline of citrus. Plant Disease Reptr. 37:379–83.

321. Suit, R. F., and E. P. DuCharme. 1967. Spreading decline of citrus. Florida Dept. Agr., Div. Plant Ind. Bull. 7.

322. Suit, R. F., and E. P. DuCharme. 1971. Cause and control of pink pitting on grapefruit. Plant Disease Reptr. 55:923–26.

323. Swingle, W. T., and H. J. Webber. 1896. The principal diseases of citrous fruits in Florida. U.S. Dept. Agr., Div. Veg. Phys. & Path. Bull. 8.

324. Tanaka, S., and S. Yamada. 1952. Studies on the greasy spot (black melanose) of citrus. I. Confirmation of the causal fungus and its taxonomic study. Hort. Div., Nat. Tokai-Kinki Agr. Expt. Sta. Bull. 1: 1–15.

325. Tarjan, A. C. 1967. Citrus nematode found widespread in Florida. Plant Disease Reptr. 51:317.

326. Tarjan, A. C., and J. H. O'Bannon. 1969. Equipment and methods used to increase productivity of Florida citrus infected with Tylenchulus semipenetrans. Phytopathology 59:1348–49.

327. Tarjan, A. C., and J. H. O'Bannon. 1969. Observations on meadow nematodes (Pratylenchus spp.) and their relation to declines of citrus in Florida. Plant Disease Reptr. 53:683–86.

328. Thompson, W. L. 1948. Greasy spot on citrus leaves. Citrus Ind. 29(4):20–22, 26.

329. Tisdale, W. B. 1934. Diseases of lime trees. Proc. Florida State Hort. Soc. 47:123–27.

330. Tisdale, W. B. 1936. Present status of lime bark diseases. Proc. Florida State Hort. Soc. 49:148–49.

331. Tsao, P. H., and J. P. Martin. 1963. Effect of soil exchangeable hydrogen and calcium ratio and the resulting pH on infection of citrus fibrous roots by Phytophthora parasitica. Phytopathology 53:1143.

332. Turrell, F. M., V. P. Sokoloff, and L. J. Klotz. 1943. Structure and composition of citrus leaves affected with mesophyll collapse. Plant Physiol. 18:463–75.

333. Underwood, L. M. 1891. Diseases of the orange in Florida. J. Mycol. 7(1):27–36.

334. Van der Plank, J. E., G. F. Van Wyk, and O. T. Van Niekerk. 1940. Removal of sooty blotch from citrus fruits. Farming S. Africa 15:201–2.

335. Van Gundy, S. D. 1958. The life-history of the citrus nematode, Tylenchulus semipenetrans Cobb. Nematologica 3:283–94.

336. Van Gundy, S. D. 1959. The life history of Hemicycliophora arenaria Raski (Nematoda, Criconematidae). Proc. Helminthol. Soc. Wash. D.C. 26:67–72.

337. Van Gundy, S. D., and J. D. Kirkpatrick. 1964. Nature of resistance in certain citrus rootstocks to citrus nematode. Phytopathology 54:419–27.

338. Van Gundy, S. D., J. P. Martin, and P. H. Tsao. 1964. Some soil factors influencing reproduction of the citrus nematode and growth reduction of sweet orange seedlings. Phytopathology 54:294–99.

339. Van Gundy, S. D., and P. H. Tsao. 1963. Growth reduction of citrus seedlings by Fusarium solani as influenced by the citrus nematode and other soil factors. Phytopathology 53:488–89.

340. Voorhees, R. K., W. T. Long, and H. J. Reitz. 1951. Citrus investigations in the coastal area. Decline studies. Florida Agr. Expt. Sta. Ann. Rept., p. 163.

341. Wallace, J. M. 1957. Virus stain interference in relation to symptoms of psorosis disease of citrus. Hilgardia 27:223–46.

342. Wallace, J. M. 1968. Psorosis A, blind pocket, concave gum, crinkly leaf, and infectious variegation, pp. 5–15. In J. F. L. Childs (ed.), Indexing Procedures for 15 Virus Diseases of Citrus Trees. U.S. Dept. Agr., Agr. Handbook 333.

343. Wallace, J. M. 1968. Recent developments in the citrus psorosis diseases, pp. 1–9. In J. F. L. Childs (ed.), Proc. 4th Conf. Intern. Organization Citrus Virol. Gainesville: University of Florida Press.

344. Wallace, J. M. 1968. Tristeza and seedling yellows, pp. 20–27. In J. F. L. Childs (ed.), Indexing Procedures for 15 Virus Diseases of Citrus Trees. U.S. Dept. Agr., Agr. Handbook 333.

345. Wallace, J. M., and R. J. Drake. 1962. Tatter leaf, a previously undescribed virus effect on citrus. Plant Disease Reptr. 46:211–12.

346. Wallace, J. M., and R. J. Drake. 1963. New information on symptom effects and host range of the citrus tatter leaf virus. Plant Disease Reptr. 47:352–53.

347. Wallace, J. M., and R. J. Drake. 1972. Studies on recovery of citrus plants from seedling yellows and the resulting protection against reinfection, pp. 127–36. In W. C. Price (ed.), Proc. 5th Conf. Intern. Orga-

nization Citrus Virol. Gainesville: University of Florida Press.

348. Wallace, J. M., and R. J. Drake. 1972. Use of seedling-yellows recovery and protection phenomena in producing tristeza-tolerant, susceptible, scion-rootstock combinations, pp. 137–43. *In* W. C. Price (ed.), Proc. 5th Conf. Intern. Organization Citrus Virol. Gainesville: University of Florida Press.

349. Waterhouse, G. M. 1956. The genus *Phytophthora*. Commonwealth Mycol. Inst. (Kew) Misc. Pub. 12.

350. Weathers, L. G. 1965. Transmission of exocortis virus of citrus by *Cuscuta subinclusa*. Plant Disease Reptr. 49:189–90.

351. Weathers, L. G., and E. C. Calavan. 1961. Additional indicator plants for exocortis and evidence for strain differences in the virus. Phytopathology 51:262–64.

352. Weathers, L. G., and F. C. Greer, Jr. 1972. Gynura as a host for exocortis virus of citrus, pp. 95–98. *In* W. C. Price (ed.), Proc. 5th Conf. Intern. Organization Citrus Virol. Gainesville: University of Florida Press.

353. Weathers, L. G., and M. K. Harjung. 1964. Transmission of citrus viruses by dodder, *Cuscuta subinclusa*. Plant Disease Reptr. 48:102–3.

354. Weathers, L. G., M. K. Harjung, and R. G. Platt. 1965. Some effects of host nutrition on symptoms of exocortis, pp. 102–7. *In* W. C. Price (ed.), Proc. 3d Conf. Intern. Organization Citrus Virol. Gainesville: University of Florida Press.

355. Webber, I. E., and H. S. Fawcett. 1935. Comparative histology of healthy and psorosis affected tissues of *Citrus sinensis*. Hilgardia 9:71–109.

356. Weber, G. F. 1927. Thread blight, a fungous disease of plants caused by *Corticium stevensii* Burt. Florida Agr. Expt. Sta. Bull. 186:143–62.

357. Weber, G. F. 1933. Occurrence and pathogenicity of *Nematospora* spp. in Florida. Phytopathology 23:384–88.

358. Weindling, R., and H. S. Fawcett. 1936. Experiments in the control of Rhizoctonia damping-off of citrus seedlings. Hilgardia 10:1–16.

359. West, E., M. Cohen, and L. C. Knorr. 1954. Brown rot of citrus fruit on the tree in Florida. Plant Disease Reptr. 38:120–21.

360. Whiteside, J. O. 1970. Etiology and epidemiology of citrus greasy spot. Phytopathology 60:1409–14.

361. Whiteside, J. O. 1970. Factors contributing to the restricted occurrence of citrus brown rot in Florida. Plant Disease Reptr. 54:608–12.

362. Whiteside, J. O. 1970. Symptomatology of orange fruit infected by the citrus greasy spot fungus. Phytopathology 60:1859–60.

363. Whiteside, J. O. 1971. Some factors affecting the occurrence and development of foot rot on citrus trees. Phytopathology 61:1233–38.

364. Whiteside, J. O. 1972. Histopathology of citrus greasy spot and identification of the causal fungus. Phytopathology 62:260–63.

365. Whiteside, J. O., and L. C. Knorr. 1970. Mushroom root rot—an easily overlooked cause of citrus tree decline. Citrus Ind. 51(11):17–20.

366. Winston, J. R. 1921. Tear stain of citrus fruits. U.S. Dept. Agr. Bull. 924.

367. Winston, J. R. 1923. Citrus scab: its cause and control. U.S. Dept. Agr. Bull. 1118.

368. Winston, J. R. 1938. Algal fruit spot of orange. Phytopathology 28:283–86.

369. Winston, J. R., and J. J. Bowman. 1923. Bordeaux-oil emulsion. U.S. Dept. Agr. Bull. 1178.

370. Winston, J. R., J. J. Bowman, and W. J. Bach. 1925. Relative susceptibility of some rutaceous plants to attack by the citrus-scab fungus. J. Agr. Res. 30:1087–93.

371. Winston, J. R., J. J. Bowman, and W. J. Bach. 1927. Citrus melanose and its control. U.S. Dept. Agr. Bull. 1474.

372. Wolf, F. A. 1926. The perfect stage of the fungus which causes melanose of citrus. J. Agr. Res. 33:621–25.

373. Wolf, F. A. 1930. A parasitic alga, *Cephaleuros virescens* Kunze, on citrus and certain other plants. J. Elisha Mitchell Sci. Soc. 45:187–205.

374. Wolf, F. A., and W. J. Bach. 1927. The thread blight disease caused by *Corticium koleroga* (Cooke) Hohn., on citrus and pomaceous plants. Phytopathology 17:689–709.

375. Wollenweber, H. W. 1931. Fusarium—Monographie fungi parasitici et saprophytici. Zeits. Wiss. Biol., Abt. B, Bd. 3, Heft 3, pp. 269–516.

376. Yamada, S. 1961. Epidemiological studies on the scab disease of Satsuma orange caused by *Elsinoe fawcetti* Bitancourt & Jenkins and its control. Tokai-Kinki Agr. Expt. Sta. Hort. Sta. Spec. Bull. 2.

377. Yamada, S., and S. Yamamoto. 1961. Studies on the epidemiology of citrus melanose and stem-end rot caused by *Diaporthe citri* (Fawc.) Wolf. Nat. Tokai-Kinki Agr. Expt. Sta. Bull. 6:108–16.

APPENDIX 1

CITRUS DISEASES AND DISORDERS OCCURRING ELSEWHERE THAN IN FLORIDA

Adjectives may be the enemies of nouns but unqualified terms in tables may be as bad or worse, posing as they often do as unembellished facts. A few prefatory remarks to the following tabulation are in order.

The column headed "Disease or Disorder" contains names that have been published in the literature. Undoubtedly, some coinages are repetitious, but until the diseases described in separate parts of the world can be compared in a common environment, no reduction to synonymy is possible. In some cases, as with Trabut's infectious chlorosis described in 1913, direct comparison with presently known diseases can no longer be made. In a few instances, entomological problems have been included (e.g., concentric ring blotch); this has been done whenever in the past the problem was thought to be pathological.

Under "Causal Agent" is given the diagnosis reported by the investigator. Where doubt exists in the mind of the reporter—or at times in the opinion of the compiler—interrogation marks are used. The absence of such marks, however, does not necessarily certify that the diagnosis is correct, for in science persistent investigation often leads to a new and closer approximation of reality.

Under "Importance" is given the compiler's estimate of whether a disease is of major or minor economic con-sequence. The number of papers published on a particular disease enters into the calculation.

Under "Distribution" are listed the countries having reported the diseases in question. A listing may indicate, on the one hand, that the disease has been sought and found; the lack of a listing, on the other, may indicate either that on investigation the disease has not been found or, alternatively, that a search has never been made. Thus, for example, lemon sieve-tube necrosis, while presently described only from California, may on investigation probably be found wherever lemons are grown.

Under "Selected References" are included various types of citations: definitive papers on the subject, reports presenting latest control measures, manuals well-illustrated for diagnostic assistance, reports justifying geographic distribution records, and publications containing extensive bibliographies.

When the number of maladies in the table is added to the number present in Florida (as discussed in the text proper), it will be seen that citrus is afflicted by more than 250 diseases and disorders. Fortunately for the grower and consumer, only a few ever attack an individual tree during its lifetime.

Disease or Disorder	Causal Agent	Importance	Parts Attacked	Distribution	Selected References[1]
Acorn disease. See Stubborn					
Adustiosis. See Red blotch					
Albedo browning	Physiological?	Minor	Fruit	USA (California)	22
Alkali injury	Chemical	Major	Tree	USA (California)	22
Areolate leaf spot	Fungal: *Corticium areolatum* Stahel	Minor	Leaves, twigs	Argentina, Brazil, Surinam	2, 6, 22, 60
Armillaria root rot	Fungal: *Armillaria mellea* (Vahl) Kummer[2]	Major	Trunks, roots	Australia, Cyprus, France (Corsica), Italy, Malawi, Malta, Tunisia, USA (California)	22, 38, 44
Ascochyta bark blotch	Fungal: *Ascochyta corticola* McAlp.	Minor	Fruit, leaves, branches	Australia, New Zealand	25
Ascochyta leaf spots	Fungal: *Ascochyta citri* Penz.	Minor	Leaves, twigs	Brazil, Costa Rica, Cyprus, Fiji, Greece, India, Italy, Mexico, Morocco, Puerto Rico, Solomon Islands, Venezuela, West Irian	22, 114
	A. hesperidearum Penz.	Minor	Leaves, twigs	Italy, Morocco	114
Australian citrus scab. See Tryon's scab					
Bacterial leaf spot of sour orange	Bacterial	Minor	Leaves	Argentina, Hong Kong,* Uruguay	25
Bacterial spot of South Africa	Bacterial: *Erwinia citrimaculans* (Doidge) Magrou	Minor	Fruit, twigs	Italy (Sicily), South Africa	22
Bark rot	Viral?	Major	Branches, trunk	China, Indonesia, Japan, Philippines	22, 49
Bark rot of sour-orange rootstock	Viral?	Minor	Crown	USA (California)	25
Black mold rots. See Rhizopus rots					
Black pit. See Blast					
Black root	Fungal: *Rosellinia bunodes* (Berk. & Br.) Sacc.	Major	Crown, roots	Argentina, Azores, Balearic Islands, Caribbean Islands, Chile, India, Indonesia, Italy, Japan, Malaysia, Mexico, Portugal, Spain, Sri Lanka	22, 68, 114

Disease or Disorder	Causal Agent	Importance	Parts Attacked	Distribution	Selected References[1]
Black spot	Fungal: *Guignardia citricarpa* Kiely	Major	Fruit, leaves	Argentina,* Australia, Brazil,* "British East Africa," Caribbean Islands,* Central America,* China, Fiji,* Hawaii,* Hong Kong, India,* Indonesia, Iran, Italy,* Japan, Malaysia,* Mozambique, New Zealand,* Nigeria, Okinawa,* Pakistan,* Peru,* Philippines,* Portugal,* Singapore,* Spain,* Rhodesia, South Africa, Sri Lanka,* Swaziland, Taiwan,* Thailand,* United Arab Republic,* USSR, Viet Nam*	37, 44, 57
Blast and black pit	Bacterial: *Pseudomonas syringae* van Hall[2]	Minor	Fruit, leaves, twigs	Australia, Cyprus, Greece, Israel, Italy, Japan, New Zealand, South Africa, Sri Lanka, Tunisia, USA (California), USSR	18, 22, 25,44
Blotchy mottle. See Greening					
Boehmeriae brown rot	Fungal: *Phytophthora boehmeriae* Sawada	Minor	Fruit	Argentina, Taiwan	29, 114
Botrytis gummosis and fruit rot	Fungal: *Botrytis cinerea* Pers. ex Fr.[2]	Minor	Fruit, twigs, trunk	Argentina, Australia, Italy, New Zealand, "Palestine," Rhodesia, South Africa, USA (California), USSR	22, 25, 102
Branch blight of grapefruit	Viral/fungal/physiological?	Minor	Branches	USA (California)	25
Brown root	Fungal: *Fomes noxius* Corner	Minor	Roots	New Guinea	114
Brown spot	Genetical/physiological?	Minor	Fruit	USA (California)	101
Budshoot wilt	Meteorological?	Minor	Budshoots	USA (California)	25
Bud-union constriction disorder of grapefruit on sour orange	Viral?	Minor	Bud union, tree	Israel	75
Calamondin bud-union disorder	Incompatibility	Minor	Bud union, tree	Brazil, South Africa, USA (California, Texas)	70
Cancrosis A. See Canker					
Cancrosis B	Bacterial: *Xanthomonas citri* var.?	Major	Fruit, leaves, twigs	Argentina, Brazil, Paraguay, Uruguay	20, 44, 62, 63

APPENDIX 1—*Continued*

Disease or Disorder	Causal Agent	Importance	Parts Attacked	Distribution	Selected References[1]
Canker	Bacterial: *Xanthomonas citri* (Hasse) Dowson	Major	Fruit, leaves, twigs, roots	Afghanistan, Andaman Islands, Assam, Brazil, Burma, Cambodia, Caroline Islands, China, Congo, Hong Kong, India, Indonesia, Ivory Coast, Japan, Korea, Liu-Kiu Islands, Malaysia, Marianna Islands, Mauritius, Nepal, New Guinea, Pakistan, Papua, Paraguay, Philippines, Reunion, Rodriguez Island, Ryukyus, Seychelles, Singapore, Sri Lanka, Taiwan, Thailand, Uruguay, USA (Guam, Hawaii), Viet Nam	4, 22, 38, 44, 62,63
Cercospora leaf spots	Fungal: *Cercospora angolensis* Carv. & Mendes	Minor	Leaves	Mozambique, West Africa	2, 114
	C. aurantia Heald & Wolfe	Minor	Leaves	USA (Mississippi, Texas)	2, 114
	C. penzigii Sacc.	Minor	Leaves	Dominican Republic*	2, 114
Charcoal rot	Fungal	Minor	Fruit	West Indies	68
Chocolate spot	Fungal: *Alternaria/Cladosporium/Penicillium* complex?	Minor	Fruit, leaves	Egypt	22
Chronic decline	Viral?	Minor	Tree	USA (California)	91
Chrysosis	Viral?	Minor	Leaves	Brazil	23
Citri brown rot	Fungal: *Phytophthora citri* Ven.	Minor	Fruit	India	114
Citri powdery mildew	Fungal: *Oidium citri*	Minor	Leaves	Brazil	22
Citricola brown rot	Fungal: *Phytophthora citricola* Saw.	Minor	Fruit	Antilles, South Africa, Taiwan	114
Citrus decline of India. See Greening (in part)					
Citrus ringspot	Viral	Minor	Leaves	Central America, Mediterranean area, USA (California)	40, 100, 113
Cobweb. See Pink disease					
Concentric ring blotch	Entomological: *Calacarus citrifolii* Keifer	Minor	Leaves	Rhodesia, South Africa	19, 22

Disease or Disorder	Causal Agent	Importance	Parts Attacked	Distribution	Selected References[1]
Convex gum	Viral?	Minor	Bark	China (Fukien)	51
Corticium blight	Fungal: *Corticium solani* Prill. & Del.?	Minor	Seedlings	Panama	25
Corynespora leaf spot	Fungal: *Corynespora citricola* M. B. Ellis	Minor	Leaves	Australia	1
Cotton root rot	Fungal: *Phymatotrichum omnivorum* (Shear) Dug.	Minor	Roots	USA (Arizona, California, Texas)	22, 25
Cottony rot. See Sclerotinia rot					
Crazy top. See Stubborn					
Cristacortis	Viral	Major	Trunk	Mediterranean area	40, 110
Crotch disease of tangerine	Viral?	Minor	Bark	Argentina, Paraguay, Uruguay	24, 25
Crown gall	Bacterial: *Agrobacterium tumefaciens* (E. F. Sm. & Town.) Conn[2],[6]	Minor	Crown	USA (Arizona, California, Texas)	2
Curvularia dieback	Fungal: *Curvularia tuberculata* Jain	Minor	Twigs	India	13, 50
Cyclosis	Viral?	Minor	Leaves	Brazil	25
Cytosporina spot	Fungal: *Cytosporina citriperda* Camp.	Minor	Fruit	Italy	12, 25
Didymella bark spot	Fungal: *Didymella citri*	Minor	Bark	Brazil	22, 64
Dothiorella gummosis	Fungal: *Dothiorella ribis* Gross. & Dug.?[3]	Minor	Fruit, limbs, trunk	Italy, USA (California)	22, 39
Dry bark of lemon	Fungal?	Minor	Branches, trunk	USA (California)	22
Dry black rot of fruit. See Pleospora rot and spot					
Dry root rot	Fungal?	Minor	Roots	Australia, Cuba, Italy, South America, USA (California), USSR	22, 25
Dweet mottle	Viral	Minor	Leaves	USA (California)	40
Ellendale decline	Viral?	Minor	Tree	Australia, South Africa	54
Epidemic dieback of lime	Viral?	Major	Roots	Trinidad	53

APPENDIX 1—*Continued*

Disease or Disorder	Causal Agent	Importance	Parts Attacked	Distribution	Selected References[1]
Eruptive bud-union crease	Viral?	Major	Tree	Egypt, India	3, 13, 66
Eruptive gummosis	Viral?	Minor	Fruit, bark	Argentina	73
Exobasidium disease	Fungal: *Exobasidium citri* Siem.	Minor	Fruit	USSR	22
Exosporina branch blight. See Hendersonula branch wilt					
False canker. See Cancrosis B					
False exanthema. See Variola					
Finger mark	Physiological?	Minor	Branches	Brazil, Italy, Turkey	58, 99
Foam disease	Fungal?	Minor	Bark	Assam	16
Frolich's Rangpur-lime disorder	Viral?	Minor	Tree	USA (California)	30
Fusarium brown rot	Fungal: *Fusarium* spp.	Minor	Fruit	Australia, India, South Africa, Trinidad, USA (California, Texas)	38, 102
Fusarium twig disease	Fungal: *Fusarium solani* Mart.	Minor	Leaves, twigs	Egypt, Honduras, South Africa	25
Gray mold. See Botrytis gummosis and fruit rot					
Greening	Mycoplasma-like organism	Major	Fruit, seed, leaves, tree	China, India, Indonesia, Malagasy Republic, Mauritius, Nepal, Pakistan, Philippines, Reunion, Rhodesia, South Africa, Swaziland, Taiwan, Thailand	28, 52, 55, 56, 84, 95, 97, 104
Gummy bark of sweet orange	Viral	Minor	Trunk	Egypt, Saudi Arabia, Sudan, USA (Florida?)	65, 67
Gum pocket	Viral?	Minor	Trifoliate portion of trunk	South Africa	98
Hard root rot	Fungal: *Rhizoctonia lamellifera* Small	Minor	Roots, trunk	Rhodesia	22, 25, 34
Hassaku dwarf	Viral	Major	Leaves, bark	Japan	88, 106
Helicobasidium felt	Fungal: *Helicobasidium mompa* Tanaka	Minor	Roots, trunk	China, Japan, Korea, Taiwan	114
Hendersonula branch wilt	Fungal: *Hendersonula toruloidea* Nattrass	Minor	Branches, trunk	USA (California)	38, 39

Disease or Disorder	Causal Agent	Importance	Parts Attacked	Distribution	Selected References[1]
Hibernalis brown rot	Fungal: *Phytophthora hibernalis* Carne	Minor	Fruit, leaves	Australia, Israel, Italy, New Zealand, Portugal, South Africa, USA (California)	22
Hyaline rot	Fungal: *Phytophthora* sp.?	Minor	Fruit	Japan	22
Impietratura	Viral	Major	Fruit	Mediterranean area, Venezuela	15, 44, 71, 81, 83, 89
Infectious mottling (Petri's)	Viral	Minor	Leaves, tree	Italy	22, 25
Knobby bark	Viral?	Minor	Bark	USA (California)	25
Leaf curl	Viral	Minor	Leaves, sprouts, limbs, trunk	Brazil	40, 85, 86
Leaf mottle yellows disease. See Greening					
Leaf mottling disease. See Greening					
Lemon/citrange bud-union degeneration	Physiological?	Minor	Tree	South Africa, USA (California)	118
Lemon sieve-tube necrosis	Genetical	Major	Tree	USA (California)	92, 93, 94
Lemon tree collapse	Viral?	Major	Tree	USA (California)	9
Leptosphaeria leaf spot	Fungal: *Leptosphaeria bondari* Bitanc. & Jenkins	Minor	Leaves	USA (Puerto Rico)	2
Likubin. See Greening					
Lime ringspot	Viral?	Minor	Leaves	Argentina, Venezuela	25, 45
Little leaf. See Stubborn					
Macrophomina root rot	Fungal: *Macrophomina phaseoli* (Maubl.) Ashby[2]	Minor	Roots	"Palestine," Rhodesia, Sri Lanka, Tanzania, USA (Arizona, California), West Indies	22
Mal secco	Fungal: *Deuterophoma tracheiphila* Petri	Major	Leaves, twigs, branches	Algeria, Colombia, Crete, Cyprus, Egypt, France, Greece, Israel, Italy, Lebanon, Syria, Tunisia, Turkey, Uganda, USSR	38, 44, 72, 114

APPENDIX 1—*Continued*

Disease or Disorder	Causal Agent	Importance	Parts Attacked	Distribution	Selected References[1]
Megasperma brown rot	Fungal: *Phytophthora megasperma* Drechs.	Minor	Roots, trunk	USA (California)	22
Membranous stain	Physiological	Minor	Fruit	USA (California)	22
Mildew. See Citri powdery mildew and Tingitaninum powdery mildew					
Mistletoe	Angiospermal: *Dendropemon* spp.	Minor	Twigs, branches	Dominica, USA (Puerto Rico)	22
	Angiospermal: *Loranthus* spp.	Minor	Twigs, branches	China, Indonesia, Mexico, Philippines, Rhodesia, Sri Lanka, Thailand, USA (Puerto Rico), Venezuela	22, 45
	Angiospermal: *Struthanthus dichotrianthus*	Minor	Twigs, branches	Trinidad	22
Mucor fruit rot	Fungal: *Mucor paronychia* Suth.-Camp. & Plunkett and *M. racemosus* Fres.	Minor	Fruit	USA (California)	2
Multiple sprouting disease	Viral?	Minor	Leaves, sprouts, stock	Rhodesia, South Africa	96
Mycosphaerella leaf spots	Fungal: *Mycosphaerella citricola* Tul.	Minor	Fruit, leaves, twigs	India	114
	M. gibellina Pass.	Minor	Fruit, leaves, twigs	USSR, Western Europe	114
	M. inflata Penz.	Minor	Fruit, leaves, twigs	Italy	114
	M. lageniformis Rehm and *M. loefgreni* Noack	Minor	Fruit, leaves, twigs	Argentina, Brazil, Dominican Republic, USA (California)	22, 74, 114
Narrow leaf	Viral	Minor	Leaves	Italy (Sardinia)	61
Navel-orange scab. See Sweet-orange scab					
Oberholzer's stem-pitting disease. See Stem-pitting disease of grapefruit					
Ombrosis	Viral?	Minor	Leaves	Brazil	25
Omphalia leaf spot	Fungal: *Omphalia flavida* (Cke.) Maubl. & Rangel	Minor	Leaves	USA (Puerto Rico)	2

Disease or Disorder	Causal Agent	Importance	Parts Attacked	Distribution	Selected References[1]
Ovularia rind spot	Fungal: Ovularia citri Briosi & Farneti	Minor	Fruit	Italy	114
Palmivora brown rot	Fungal: Phytophthora palmivora Butler	Major	Fruit, leaves, stems, trunk	India, Indonesia (Java), Malaysia, Philippines, Sri Lanka, Surinam, Tanzania, Trinidad, USA (Puerto Rico)	22
Pestalotiopsis fruit spot	Fungal: Pestalotiopsis neglecta (Thuem.) Steg.	Minor	Fruit	Guinea	114
Peteca	Physiological: suboxidation	Minor	Fruit	Italy, USA (California)	2, 22, 38
Petrifaction. See Impietratura					
Phoma rot	Fungal: Phoma citricarpa McAlp. var. mikan Hara	Minor	Fruit	Japan	114
Phyllosticta spots	Fungal: Phyllosticta apiculata Sacc. & Syd.	Minor	Leaves	Italy	114
	P. arethusa Bub.	Minor	Leaves	Australia, Italy	114
	P. auranticola (Berk. & Curt.) Sacc.	Minor	Leaves	Colombia, Dominican Republic	114
	P. beltranii Penz.	Minor	Leaves	Italy	114
	P. circumsepta Sacc.	Minor	Fruit	Philippines	114
	P. citricola Hori	Minor	Leaves, roots	Japan, "Palestine," USA (Mississippi)	31
	P. curvispora Hori	Minor	Leaves	Japan, Taiwan	114
	P. deliciosa Pass.	Minor	Leaves	Italy	114
	P. disciformis Penz. var. brasiliensis Speg.	Minor	Leaves	Brazil, Italy, Sri Lanka, USSR	114
	P. fuliginosa Massal.	Minor	Leaves	Italy	114
	P. lenticularis Pass.	Minor	Leaves	Portugal	114
	P. micrococcoides Penz.	Minor	Leaves	Italy	114
Pink disease	Fungal: Corticium salmonicolor Berk. & Br.[2]	Major	Twigs, branches, trunk	Antilles, Australia, Brazil, Cameroons, East Indies, India, Japan, Kenya, Nepal, Philippines, Sri Lanka, Taiwan, Thailand, USA (Puerto Rico), USSR	22, 44

APPENDIX 1—Continued

Disease or Disorder	Causal Agent	Importance	Parts Attacked	Distribution	Selected References[1]
Pink mold	Fungal: *Penicillium roseum* Link[2]	Minor	Fruit	USA (California)	21
Pink nose. See Stubborn					
Pleospora rot and spot	Fungal: *Pleospora herbarum* (Pers.) Raben. and *P.* spp.	Minor	Fruit, leaves	Australia, Italy, Tunisia, USA (California)	22, 25
Pop corn	?	Minor	Bark	Brazil, Italy	99
Pyrenochaeta leaf spot	Fungal: *Pyrenochaeta destructiva* McAlp.	Minor	Leaves	Australia	114
Ramularia leaf spots	Fungal: *Ramularia citri* Penz.	Minor	Leaves	Italy	114
	R. citrifolia Saw.	Minor	Leaves	Taiwan	114
Red blotch	Physiological?	Minor	Fruit	South Africa, USA (California)	22
Red root rot	Fungal: *Sphaerostilbe repens* Berk. & Br.[5]	Major	Roots	Dominica, Ghana, Guyana	22
Rhizopus rots	Fungal: *Rhizopus nigricans* Ehr. and *R. stolonifer* (Ehr. ex Fr.) Lind	Minor	Fruit	Cuba, USA (California, Puerto Rico)	2, 32
Rind stipple of grapefruit	Meteorological?	Minor	Fruit	USA (California)	41
Rubellosis. See Pink disease					
Satsuma anthracnose	Fungal: *Gloeosporium folicolum* Nishida	Minor	Fruit	Japan	33
Satsuma dwarf	Viral	Major	Tree	Japan	105, 107
Sclerotinia rot	Fungal: *Sclerotinia sclerotiorum* (Lib.) deB.[2]	Minor	Fruit, twigs, branches, trunk, roots	Argentina, Australia, Brazil, Italy, Japan, New Zealand, "Palestine," USA (California, Texas)	22
Septoria spots	Fungal: *Septoria arethusa* Penz.	Minor	?	Argentina, Brazil, Italy, Morocco	114
	S. cattanei Thuem.	Minor	?	Canary Islands, Europe, India	114
	S. cinerescens (Dur. & Mont.) Sacc.	Minor	Leaves	Algeria, Europe	114

Disease or Disorder	Causal Agent	Importance	Parts Attacked	Distribution	Selected References[1]
	S. citri Pass.	Minor	Fruit, leaves	Argentina, Australia, Brazil, Chile, Cyprus, El Salvador, Greece, Israel, Italy, Mexico, New Zealand, North Africa, Singapore,* South Africa, Spain, USA (California)	22, 114
	S. depressa McAlp.	Minor	Fruit	Australia, Europe	114
	S. flaccescens McAlp.	Minor	Leaves	Australia	114
	S. glaucescens Trab.	Minor	?	Algeria, Morocco, Tunisia	114
	S. limonum Pass.	Minor	Fruit	USA (California)	22, 25
	S. sicula Penz.	Minor	Leaves	Europe	114
	S. westraliensis McAlp.	Minor	Leaves	Australia, Europe	114
Siccortosis. See Dry bark of lemon					
Small fruit of grapefruit	?	Minor	Fruit	Argentina	26
Smoky blotch	Fungal: Stemiopeltis citri	Minor	Fruit	Southern Rhodesia	25
Sour-orange rootstock necrosis	Genetical?	Minor	Tree	USA (California)	90
Spongium felt	Fungal: Septobasidium spongium (Berk. & Curt.) Pat.	Minor	Twigs	USA (Puerto Rico)	2
Stem-pitting disease of grapefruit	Viral (tristeza complex)[4]	Major	Fruit, branches, trunk, tree	Aden, Argentina, Australia, Brazil, Congo, India, Japan, Mozambique, South Africa, Swaziland, Uruguay	36, 42, 44, 46, 69, 77, 103
Stubborn	Mycoplasma-like organism	Major	Fruit, seed, leaves, tree	Possibly worldwide	10, 11, 14, 15, 38, 40, 44, 76, 84
Stunting of grapefruit trees	?	Minor	Tree	Argentina	26
Sudden death	Fungal: Coprinus micaceus?	Minor	Roots	Australia	59
Syringae brown rot	Fungal: Phytophthora syringae Kleb.	Minor	Fruit	USA (California)	22
Sweet-orange scab	Fungal: Elsinoë australis Bitanc. & Jenkins	Major	Fruit, leaves, twigs	Argentina, Bolivia, Brazil, Dominican Republic,* Ecuador,* Italy (Sicily), New Caledonia, Paraguay, Uruguay	7, 8, 43, 44

APPENDIX 1—Continued

Disease or Disorder	Causal Agent	Importance	Parts Attacked	Distribution	Selected References[1]
Sweet-orange verrucosis. See Sweet-orange scab					
Tarocco pit	Viral?	Major	Fruit, shoots, leaves, trunk	Italy	82
Tingitaninum powdery mildew	Fungal: *Oidium tingitaninum* Carter	Minor	Leaves	India, Indonesia (Java), Nepal, Sri Lanka, USA (California)	22, 47
Trabut's infectious chlorosis	Viral?	Minor	Leaves	Algeria	108
Tracheoverticillosis	Fungal: *Verticillium albo-atrum* Reinke & Berth.?[2]	Minor	Roots	Italy	80
Trametes wood rot	Fungal: *Trametes hispida* Bagl. and *T. hydnoides* Sw. ex Fr.	Minor	Wood	USA (California)	2
Trichoderma rot	Fungal: *Trichoderma viride* Pers. ex Fr.	Major	Fruit, roots	USA (California) and probably cosmopolitan in cool situations	22, 25, 102
Tryblidiella twig blight	Fungal: *Tryblidiella rufula* (Spreng.) Sacc.	Minor	Twigs	USA (Puerto Rico, Texas)	2
Tryon's scab	Fungal: *Sphaceloma fawcettii* Jenkins var. *scabiosa* (McAlp. & Tryon) Jenkins	Major	Leaves	Argentina, Australia, New Caledonia, New Guinea, New Zealand, Rhodesia, South Africa, Sri Lanka	22, 35, 114
Ultimum rootlet rot	Fungal: *Pythium ultimum* Trow	Minor	Rootlets	USA (California)	2
Uredo rust	Fungal: *Uredo citri* Vaheeduddin	Minor	Leaves	India	114
Uromyces rust	Fungal: *Uromyces nilagiricus* Ramak.	Minor	Leaves	India	114
Valencia rind spot	Meteorological?	Minor	Fruit	USA (California)	25
Vaporaria root rot	Fungal: *Poria vaporaria* (Pers. ex Fr.) Cke.	Minor	Roots	USA (California)	2
Variola	Viral/entomological?	Minor	Fruit, leaves, twigs, branches	Brazil	87
Vasudeva's viral die-back	Viral?	Minor	Tree	India	109
Vein enation. See Woody gall					
Verruga. See Sweet-orange scab					

Disease or Disorder	Causal Agent	Importance	Parts Attacked	Distribution	Selected References[1]
Water spot	Meteorological	Minor	Fruit	USA (California)	25
Watson's citrus tree decline	Viral?	Minor	Tree	Iraq	115
White root rot	Fungal: *Rosellinia pepo* Pat.	Major	Roots	Caribbean Islands	22, 68, 114
Woody gall	Viral	Minor	Leaves, trunk	Australia, Japan, Peru, South Africa, USA (California)	17, 27, 38, 44, 48, 111, 112
Yellow shoot. See Greening					
Yellow vein	Viral	Minor	Leaves, tree	USA (California)	116, 117
Zonate chlorosis	Viral/entomological?	Minor	Fruit, leaves	Brazil, Surinam	5, 22, 25, 78, 79

* Countries believed to have been the origin of fruit shipments in which port-of-entry quarantine inspectors detected the causal agent.
1. Numbers relate to the bibliography at the end of this table.
2. This organism, though present in Florida on various hosts, has not been reported in the state on citrus.
3. This organism has been reported to cause a fruit rot in Florida but not a bark rot.
4. Though tristeza virus is known to occur in Florida, the stem-pitting symptom as it affects grapefruit trees has not been reported (see discussion under **Tristeza** in the text proper).
5. Recent information suggests that the disease results from poor water relations and that the fungus is merely a secondary invader.
6. Inoculation under field conditions doubtful.

APPENDIX 1 — SELECTED REFERENCES

1. Anonymous. 1958. Report of the Department of Agriculture, New South Wales, for the year ended 30 June, 1957.

2. Anonymous. 1960. Index of Plant Diseases in the United States. U.S. Dept. Agr., Agr. Handbook 165.

3. Bhutani, V. P., J. C. Bakhshi, and L. C. Knorr. 1972. Biochemical changes in healthy and decline sweet orange trees associated with bud-union crease, pp. 229–33. In W. C. Price (ed.), Proc. 5th Conf. Intern. Organization Citrus Virol. Gainesville: University of Florida Press.

4. Bitancourt, A. A. 1957. O cancro citrico. O Biologico 23:101–11.

5. Bitancourt, A. A., and H. V. S. Grillo. 1934. A clorose zonada, nova doenca dos Citrus. Arquiv. Inst. Biol. (Sao Paulo) 5:245–50.

6. Bitancourt, A. A., and A. E. Jenkins. 1935. Areolate spot of citrus caused by Leptosphaeria bondari. Phytopathology 25:884–86.

7. Bitancourt, A. A., and A. E. Jenkins. 1936. Perfect state of the sweet orange fruit scab fungus. Mycologia 28:489–92.

8. Bitancourt, A. A., and A. E. Jenkins. 1937. Sweet orange fruit scab caused by Elsinoe australis. J. Agr. Res. 54:1–18.

9. Calavan, E. C. 1949. Lemon tree collapse. Phytopathology 39:858–59.

10. Calavan, E. C. 1968. Stubborn, pp. 35–43. In J. F. L. Childs (ed.), Indexing Procedures for 15 Virus Diseases of Citrus Trees. U.S. Dept. Agr., Agr. Handbook 333.

11. Calavan, E. C., and J. B. Carpenter. 1965. Stubborn disease retards growth, impairs quality, and decreases yield. Calif. Citrograph 50:86–87, 96, 98–99.

12. Campanile, G. 1922. In di una malettia della frutta di mandarino. Le Staz. Sper. Agr. Ital. 55:5–12.

13. Chadha, K. L., N. S. Randhawa, O. S. Bindra, J. S. Chohan, and L. C. Knorr. 1970. Citrus Decline in India—Causes and Control. Ludhiana, Punjab: Punjab Agr. Univ./Ohio State Univ./U.S. Agency Intern. Devel.

14. Chapot, H., J. Cassin, and M. Larue. 1962. Nouvelles varietes d'agrumes atteintes par le stubborn. Al Awamia (Morocco) 4:1–6.

15. Chapot, H., and V. L. Delucchi. 1964. Maladies, Troubles, et Ravageurs des Agrumes au Maroc. Rabat, Morocco: Institut National de la Recherche Agronomique.

16. Chowdhury, S. 1950. Foam disease of citrus in Assam. Current Sci. (India) 19(2):62–63.

17. Desjardins, P. R., C. Fuertes-Polo, and J. M. Wallace. 1968. Tissue culture studies of West Indian lime infected with vein-enation virus, pp. 216–21. In J. F. L. Childs (ed.), Proc. 4th Conf. Intern. Organization Citrus Virol. Gainesville: University of Florida Press.

18. DeWolfe, T. A., H. C. Meith, A. O. Paulus, F. Shibuya, L. J. Klotz, R. B. Jeter, and M. J. Garber. 1966. Control of citrus blast in northern California. Calif. Agr. 20(8):12–13.

19. Dippenaar, B. J. 1958. Concentric ring blotch of citrus—its cause and control. S. African J. Agr. Sci. 1:83–108.

20. DuCharme, E. P. 1950. La causa de la cancrosis del limon. Idia 3(33–34):27–28.

21. Fawcett, H. S. 1925. The decay of citrus fruits on arrival and in storage at eastern markets. Calif. Citrograph 10:79, 98, 103.

22. Fawcett, H. S. 1936. Citrus Diseases and Their Control. N.Y.: McGraw-Hill.

23. Fawcett, H. S., and A. A. Bitancourt. 1937. Relatorio sobre as doencas dos citrus nos estados de Pernambuco, Bahia, Sao Paulo e Rio Grande do Sul. Rodriguesia 3:214–36.

24. Fawcett, H. S., and A. A. Bitancourt. 1940. Observaciones sobre las enfermedades de los citrus en el Uruguay. Asoc. Ingen. Agron. Rev. 12(3):3–8.

25. Fawcett, H. S., and L. J. Klotz. 1948. Diseases and their control, pp. 495–596. In L. D. Batchelor and H. J. Webber (eds.), The Citrus Industry, vol. 2. Berkeley: University of California Press.

26. Fernandez Valiela, M. V. 1968. Small fruit and stunting, two new disorders of grapefruit trees in the Delta del Parana and San Pedro areas of Argentina, pp. 213–15. In J. F. L. Childs (ed.), Proc. 4th Conf. Intern. Organization Citrus Virol. Gainesville: University of Florida Press.

27. Fraser, L. R. 1958. Virus diseases of citrus in Australia.

Proc. Linnean Soc. N. S. Wales 73(1):9–19.

28. Fraser, L. R., D. Singh, S. P. Capoor, and T. K. Nariani. 1966. Greening virus, the likely cause of citrus dieback in India. F.A.O. Plant Protect. Bull. 14:127–30.

29. Frezzi, M. J. 1950. Las especies de *Phytophthora* en la Argentina. Rev. Invest. Min. Agr. 4:47–134.

30. Frolich, E. F. 1958. A disorder of Rangpur lime on sweet orange rootstock. Plant Disease Reptr. 42:500–501.

31. Hara, K. 1917. Small, brown, round spot disease. J. Hort. Soc. Japan 29(9).

32. Harter, L. L., and J. L. Weimer. 1922. Decay of various vegetables and fruits by different species of *Rhizopus*. Phytopathology 12:205–12.

33. Hemmi, T. 1921. On the pathogenic nature of Nishida's anthracnose fungus of citrus. Japan. J. Plant Protect. 8:173–77.

34. Hopkins, J. C. F. 1929. Investigations into "collar-rot" disease of citrus. Rhodesia Agr. 26:137–46.

35. Jenkins, A. E., L. C. Knorr, and A. A. Bitancourt. 1953. Notes on spot anthracnoses and related subjects. V. Discovery of Tryon's citrus scab in Argentina. Rev. Arg. Agron. 20:230–32.

36. Kawata, S., and F. Ikeda. 1970. Morphogenetical studies on stem pitting in citrus trees. III. Secondary xylem formation in the stem. J. Japan. Soc. Hort. Sci. 39:21–31.

37. Kiely, T. B. 1970. New fungicidal spray programmes for coastal Valencia oranges—their effect on black spot control and fruit quality. Sci. Bull. Dept. Agr. N. S. Wales 80.

38. Klotz, L. J. 1961. Color Handbook of Citrus Diseases. Berkeley: University of California Division of Agricultural Sciences.

39. Klotz, L. J., and E. C. Calavan. 1969. Gum diseases of citrus in California. Calif. Agr. Expt. Sta. Circ. 396.

40. Klotz, L. J., E. C. Calavan, and L. G. Weathers. 1972. Virus and viruslike diseases of citrus. Calif. Agr. Expt. Sta. Circ. 559.

41. Klotz, L. J., T. A. DeWolfe, and M. P. Miller. 1971. Rind stipple of grapefruit. A progress report. Calif. Citrograph 56:248, 261–62.

42. Knorr, L. C. 1956. Suscepts, indicators, and filters of tristeza virus, and some differences between tristeza in Argentina and in Florida. Phytopathology 46:557–60.

43. Knorr, L. C. 1963. Dossier on sweet-orange scab. Citrus Ind. 44(9):7, 9, 12, 26.

44. Knorr, L. C. 1965. Serious diseases of citrus foreign to Florida. Florida Dept. Agr. Div. Plant Ind. Bull. 5.

45. Knorr, L. C., G. Malaguti, D. Serpa, and F. Leal. 1964. World citrus problems. IV. Venezuela. F.A.O. Plant Protect. Bull. 12:121–28.

46. Knorr, L. C., E. C. Paterson, and J. H. Proctor. 1961. World citrus problems. I. Aden Protectorate. F.A.O.

Plant Protect. Bull. 9:91–98.

47. Knorr, L. C., and S. M. Shah. 1971. World citrus problems. V. Nepal. F.A.O. Plant Protect. Bull. 19:73–79.

48. Laird, E. F., and L. G. Weathers. 1961. *Aphis gossypii*, a vector of citrus vein-enation virus. Plant Disease Reptr. 45:877.

49. Lee, H. A. 1923. California scaly bark and bark rot of citrus trees in the Philippines. Philippine Agr. Rev. 16:219–25.

50. Lele, V. C., S. P. Raychaudhuri, R. B. Bhalla, and A. Ram. 1968. *Curvularia tuberculata*, a new fungus causing die-back disease of citrus in India. Indian Phytopathol. 21:66–72.

51. Lin, K. H. 1943. Convex gum, a new disease of citrus in China. Phytopathology 33:394–97.

52. Lin, K. H. 1964. A preliminary study of the resistance of yellow shoot virus and citrus budwood tissue to heat. Acta Phytopathol. Sinica (Peking) 7:61–65.

53. Lucie-Smith, M. N. 1951. The cultivation of West Indian limes. J. Agr. Soc. Trinidad Tobago 51 (Suppl.).

54. McClean, A. P. D. 1956. Tristeza and stem pitting diseases of citrus in South Africa. F.A.O. Plant Protect. Bull. 4:88–94.

55. McClean, A. P. D., and P. C. J. Oberholzer. 1965. Citrus psylla, a vector of the greening disease of sweet orange. S. African J. Agr. Sci. 8:287–98.

56. McClean, A. P. D., and R. E. Schwarz. 1970. Greening or blotchy-mottle disease of citrus. Phytophylactica 2:177–94.

57. McOnie, K. C., C. Kellerman, and D. J. Kruger. 1969. Benlate, a highly promising new fungicide for the control of black spot. S. African Citrus J. 423:7–9.

58. Madaluni, A. L. 1968. Studies on finger mark disorder of citrus in Italy, pp. 10–13. *In* J. F. L. Childs (ed.), Proc. 4th Conf. Intern. Organization Citrus Virol. Gainesville: University of Florida Press.

59. Magee, C. J. 1957. Death of citrus on trifoliata stock. Commonwealth Phytopathol. News 3(4):58.

60. Marchionatto, J. B. 1947. Hongos parasitos de las plantas, nuevos o poco conocidos en la Argentina. Arg. Min. Agr., Instit. San. Veg. 3(37).

61. Marras, F. 1970. "Stenofillia:" nuova malatia de virus degli agrumi, pp. 302–5. *In* Stud. Sassaresi, sez. 3, vol. 17.

62. Namekata, T. 1971. Estudos comparativos entre *Xanthomonas citri* (Hasse) Dowson, agente causal do cancro citrico e uma bacteria agente causal da cancrose do limoeiro galego, p. 60. *In* Proc. Congresso Brasileiro de Fruticultura 1. Campinas, Sao Paulo.

63. Namekata, T., and A. R. Oliveira. 1971. Comparative serological studies between *Xanthomonas citri* and a bacterium causing canker on Mexican lime, pp. 151–52. *In* Proc. 3d Intern. Conf. Plant Pathogenic Bacteria. Wageningen.

64. Noack, F. 1900. Pilzkrankheiten der orangenbaueme in Brazilian. Pflanzenkrankh. 10:321–35.

65. Nour-Eldin, F. 1956. Phloem discoloration of sweet orange. Phytopathology 46:238–39.

66. Nour-Eldin, F. 1959. Citrus virus disease research in Egypt, pp. 219–27. *In* J. M. Wallace (ed.), Citrus Virus Diseases. Riverside: University of California Division of Agricultural Sciences.

67. Nour-Eldin, F. 1968. Gummy bark of sweet orange, pp. 50–53. *In* J. F. L. Childs (ed.), Indexing Procedures for 15 Virus Diseases of Citrus Trees. U.S. Dept. Agr., Agr. Handbook 333.

68. Nowell, W. 1923. Diseases of Crop Plants in the Lesser Antilles. London: West Indian Committee.

69. Oberholzer, P. C. J., I. Mathews, and S. F. Stiemie. 1949. The decline of grapefruit trees in South Africa. A preliminary report on so-called "stem pitting." Union S. Africa Dept. Agr. Sci. Bull. 297.

70. Olson, E. O., and E. F. Frolich. 1968. Bud-union crease of calamondin—a non-infectious disorder, pp. 341–47. *In* J. F. L. Childs (ed.), Proc. 4th Conf. Intern. Organization Citrus Virol. Gainesville: University of Florida Press.

71. Papasolomontos, A. 1969. A report on impietratura disease of citrus: its distribution and importance, pp. 1457–62. *In* H. D. Chapman (ed.), Proc. 1st Intern. Citrus Symp., vol. 3. Riverside: University of California.

72. Petri, L. 1940. Recenti ricerche sul "mal secco" degli agrumi in Turchia. Boll. Staz. Pat. veg. Roma, N.S. 20(2):81–98.

73. Pujol, A. R. 1968. Eruptive gummosis, a new virus disease of citrus, pp. 193–96. *In* J. F. L. Childs (ed.), Proc. 4th Conf. Intern. Organization Citrus Virol. Gainesville: University of Florida Press.

74. Rehm, H. 1911. New fungi occurring on orange leaves. Pomona Coll. Econ. Bot. 1:106.

75. Reichert, I., A. Bental, and O. Ginsburg. 1965. Bud-union constriction disorder of grapefruit on sour orange in Israel, pp. 192–98. *In* W. C. Price (ed.), Proc. 3d Conf. Intern. Organization Citrus Virol. Gainesville: University of Florida Press.

76. Reichert, I., A. Bental, and I. Yoffe. 1956. On the problem of the identity of "little leaf" and "xyloporosis" diseases. Ktavim 6:77–82.

77. Rossetti, V., T. G. Fassa, and J. T. Nakadaira. 1965. Reaction of citrus varieties to the stem pitting virus of Pera orange, pp. 46–48. *In* W. C. Price (ed.), Proc. 3d Conf. Intern. Organization Citrus Virol. Gainesville: University of Florida Press.

78. Rossetti, V., C. C. Lasca, J. T. Nakadaira, and C. J. S. Aguiar. 1968. A preliminary report on transmission of zonate chlorosis and varietal susceptibility to the disease, pp. 347–50. *In* J. F. L. Childs (ed.), Proc. 4th Conf. Intern. Organization Citrus Virol. Gainesville: University of Florida Press.

79. Rossetti, V., C. C. Lasca, and S. Negretti. 1969. New developments regarding leprosis and zonate chlorosis of citrus, pp. 1453–56. *In* H. D. Chapman (ed.), Proc. 1st Intern. Citrus Symp., vol. 3. Riverside: University of California.

80. Ruggieri, G. 1946. Possibili casi di tracheoverticillosi fra gli agrumi. Ital. Agr. 1946(8).

81. Ruggieri, G. 1968. Impietratura, pp. 60–62. *In* J. F. L. Childs (ed.), Indexing Procedures for 15 Virus Diseases of Citrus Trees. U.S. Dept. Agr., Agr. Handbook 333.

82. Russo, F., and L. J. Klotz. 1963. Tarocco pit. Calif. Citrograph 48:221–22.

83. Safran, H. 1969. Anatomical changes in citrus with the impietratura disease. Phytopathology 59:1226–28.

84. Saglio, P., D. Lafleche, C. Bonissol, and J. M. Bove. 1971. Isolement, culture et observation au microscope electronique des structures de type mycoplasme associees a la maladie du stubborn des agrumes et leur comparaison avec les structures observees dans le cas de la maladie du greening des agrumes. Physiol. Veg. 9(4):569–82.

85. Salibe, A. A. 1959. Leaf curl—a transmissible virus disease of citrus. Plant Disease Reptr. 43:1081–83.

86. Salibe, A. A. 1968. Leaf curl, pp. 74–76. *In* J. F. L. Childs (ed.), Indexing Procedures for 15 Virus Diseases of Citrus Trees. U.S. Dept. Agr., Agr. Handbook 333.

87. Salibe, A. A., and S. Moreira. 1965. Variola—a probable virus disease of citrus, pp. 207–9. *In* W. C. Price (ed.), Proc. 3d Conf. Intern. Organization Citrus Virol. Gainesville: University of Florida Press.

88. Sasaki, A. 1966. Studies on Hassaku dwarf. I. Detection of citrus viruses in a Hassaku tree severely affected by Hassaku dwarf. Hiroshima Agr. Expt. Sta. Rept. 23:39–47.

89. Scaramuzzi, G., A. A. Catara, and G. Cartia. 1968. Investigations on impietratura disease, pp. 197–200. *In* J. F. L. Childs (ed.), Proc. 4th Conf. Intern. Organization Citrus Virol. Gainesville: University of Florida Press.

90. Schneider, H. 1956. Decline of lemon trees on sour orange rootstock. Calif. Citrograph 41:117–20.

91. Schneider, H. 1957. Chronic decline, a tristeza-like bud union disorder of orange trees. Phytopathology 47:279–84.

92. Schneider, H. 1960. Sieve-tube necrosis in nucellar lemon trees. Calif. Citrograph 45:208, 219–22.

93. Schneider, H. 1969. Pathological anatomies of citrus affected by virus diseases and by apparently-inherited disorders and their use in diagnoses, pp. 1489–94. *In* Proc. 1st Intern. Citrus Symp., vol. 3. Riverside: University of California.

94. Schneider, H., J. W. Cameron, R. K. Soost, and E. C. Calavan. 1961. Classifying certain diseases as inherited, pp. 15–21. *In* W. C. Price (ed.), Proc. 2nd Conf. Intern. Organization Citrus Virol. Gainesville: University of Florida Press.

95. Schwarz, R. E. 1968. Greening disease, pp. 87–90. *In* J. F. L. Childs (ed.), Indexing Procedures for 15 Virus

Diseases of Citrus Trees. U.S. Dept. Agr., Agr. Handbook 333.

96. Schwarz, R. E. 1970. A multiple sprouting disease of citrus. Plant Disease Reptr. 54:1003–7.

97. Schwarz, R. E. 1972. A review of stubborn and greening diseases of citrus, pp. 1–5. In W. C. Price (ed.), Proc. 5th Conf. Intern. Organization Citrus Virol. Gainesville: University of Florida Press.

98. Schwarz, R. E., and A. P. D. McClean. 1969. A new virus-like disease of Poncirus trifoliata. Plant Disease Reptr. 53:336–39.

99. Servazzi, O., F. Marras, and A. Foddai. 1968. "Finger marks" e "pop corn" degli Agrumi in Sardegna, pp. 227–34. In Stud. Sassaresi, sez. 3, vol. 16.

100. Servazzi, O., F. Marras, and A. Foddai. 1968. La "maculatura anulare" ("Ring Spot") degli agrumi in Sardegna, pp. 520–27. In Stud. Sassaresi, sez. 3, vol. 16.

101. Shamel, A. D., C. S. Pomeroy, and R. E. Caryl. 1924. Bud selection as related to quality of crop in Washington Navel orange. J. Agr. Res. 28:521–25.

102. Smoot, J. J., L. G. Houck, and H. B. Johnson. 1971. Market Diseases of Citrus and Other Subtropical Fruits. U.S. Dept. Agr., Agr. Handbook 398.

103. Steyaert, R. L. 1952. La "tristeza" des agrumes. Bull. Agr. Congo Belge 43:399–446.

104. Su, H. J., and T. Matsumoto. 1972. Further studies on the complex causing likubin of citrus in Taiwan, pp. 28–34. In W. C. Price (ed.), Proc. 5th Conf. Intern. Organization Citrus Virol. Gainesville: University of Florida Press.

105. Tanaka, S. 1968. Satsuma dwarf, pp. 56–59. In J. F. L. Childs (ed.), Indexing Procedures for 15 Virus Diseases of Citrus Trees. U.S. Dept. Agr., Agr. Handbook 333.

106. Tanaka, S., E. Shikata, and A. Sasaki. 1969. Studies on Hassaku-dwarf virus, pp. 1445–48. In Proc. 1st Intern. Citrus Symp., vol. 3. Riverside: University of California.

107. Tanaka, H., and S. Yamada. 1972. Evidence for a relationship among the viruses of Satsuma dwarf, citrus mosaic, Navel-infectious-mottling, Natsudaidai dwarf, citrus variegation, and citrus crinkly leaf, pp. 71–76. In W. C. Price (ed.), Proc. 5th Conf. Intern. Organization Citrus Virol. Gainesville: University of Florida Press.

108. Trabut, L. 1913. Chlorose infectieuse des Citrus. Compt. Rend. 156:243–44.

109. Vasudeva, R. S. 1957. News from India. Commonwealth Phytopathol. News 3(2):28–29.

110. Vogel, R., and J. M. Bove. 1972. Relation of cristacortis virus to other citrus viruses, pp. 178–84. In W. C. Price (ed.), Proc. 5th Conf. Intern. Organization Citrus Virol. Gainesville: University of Florida Press.

111. Wallace, J. M. 1968. Vein enation and woody gall, pp. 44–49. In J. F. L. Childs (ed.), Indexing Procedures for 15 Virus Diseases of Citrus Trees. U.S. Dept. Agr., Agr. Handbook 333.

112. Wallace, J. M., and R. J. Drake. 1953. New virus found in citrus. Calif. Citrograph 38:180–81.

113. Wallace, J. M., and R. J. Drake. 1968. Citrange stunt and ringspot, two previously undescribed virus diseases of citrus, pp. 177–83. In J. F. L. Childs (ed.), Proc. 4th Conf. Intern. Organization Citrus Virol. Gainesville: University of Florida Press.

114. Watson, A. J. 1971. Foreign Bacterial and Fungus Diseases of Food, Forage, and Fiber Crops: An Annotated List. U.S. Dept. Agr., Agr. Handbook 418.

115. Watson, R. D., and A. R. Al-Adhami. 1957. Notes on diseases of fruit trees in Iraq. F.A.O. Plant Protect. Bull. 5:104–7.

116. Weathers, L. G. 1957. A vein-yellowing disease of citrus caused by a graft-transmissible virus. Plant Disease Reptr. 41:741–42.

117. Weathers, L. G. 1968. Yellow vein, pp. 54–55. In J. F. L. Childs (ed.), Indexing Procedures for 15 Virus Diseases of Citrus Trees. U.S. Dept. Agr., Agr. Handbook 333.

118. Weathers, L. G., E. C. Calavan, J. M. Wallace, and D. W. Christiansen. 1955. A bud union and rootstock disorder of Troyer citrange with Eureka lemon tops. Plant Disease Reptr. 39:665–69.

APPENDIX 2

COMMON NAMES OF CITRUS VARIETIES
AND THEIR TECHNICAL EQUIVALENTS

The center column of the following tabulation lists in alphabetical order the common names of citrus varieties mentioned in the text. In the columns to the left and right are the equivalent scientific binomials according to the two major technical systems for designating species. Both systems are in current use, which often results in the unfortunate employment of two different scientific names for the same species.

According to Tanaka, whose system appears in the left-hand column, the genus *Citrus* embraces 159 species, each being distinguished by apparent horticultural differences. In contrast, Swingle (righthand column) utilizes a natural system of classification in which he regards the genus *Citrus* as containing only 16 basic species, the rest being considered simply as hybrids and cultivars.

A discussion of principles involved in the systematics of *Citrus* is given in the chapter "The Botany of Citrus and its Wild Relatives of the Orange Subfamily" in W. Reuther, H. J. Webber, and L. D. Batchelor (eds.), The Citrus Industry, 2d ed., vol. 1, Univ. of California Div. Agr. Sci., Berkeley, 1967. An extensive list of equivalent names, as prepared by Dr. P. C. Reece, may be found in "Classification of Citrus Species," pp. 91–95, U.S. Dept. Agr., Agr. Res. Serv., Agr. Handbook 333, 1968.

Gratitude for assistance in the preparation of the following list is expressed to Dr. Reece, former horticulturist at the U.S. Department of Agriculture, Orlando, Florida.

Tanaka System	Common Name	Swingle System
Citrus Tanaka		*Citrus* Swingle
macrophylla Wester	ALEMOW	*celebica* Koord. hybrid
sinensis Osbeck	ALGERIAN NAVEL SWEET ORANGE	*sinensis* (L.) Osbeck
limon (L.) Burm. f.	ARIZONA LEMON	*limon* (L.) Burm. f.
limon (L.) Burm. f.	AVON LEMON	*limon* (L.) Burm. f.
limon (L.) Burm. f.	BEARSS LEMON	*limon* (L.) Burm. f.
?	BUTWAL SWEET LIME	?
madurensis Lour.	CALAMONDIN	*reticulata* hybrid
?	CARRIZO CITRANGE	*sinensis* X *P. trifoliata*
?	CITREMON	*P. trifoliata* X *C. limon*
medica L.	CITRON	*medica* L.
limonimedia Lush.	CITRON ETROG	*medica* L. var. *ethrog* Engl.
?	CITRUMELO	*P. trifoliata* X *C. paradisi*
clementina Hort. ex Tan.	CLEMENTINE MANDARIN	*reticulata* Blanco
lycopersicaeformis Hort. ex Tan.	CLEOPATRA MANDARIN	*reticulata* Blanco
?	COLUMBIAN SWEET LIME	?
paradisi Macf.	CONNER GRAPEFRUIT	*paradisi* Macf.
?	CUBAN SHADDOCK	hybrid
tangerina Hort. ex Tan.	DANCY MANDARIN	*reticulata* Blanco
?	DOMINICAN THORNLESS LIME	*aurantifolia* (Christm.) Swing.
limon (L.) Burm. f.	DO.SA.CO. LEMON	*limon* (L.) Burm. f.
paradisi Macf.	DUNCAN GRAPEFRUIT	*paradisi* Macf.
jambhiri Lush.	ESTES ROUGH LEMON	*limon* (L.) Burm. f.
limon (L.) Burm. f.	EUREKA LEMON	*limon* (L.) Burm. f.
?	EUSTIS LIMEQUAT	*Fortunella japonica* X *C. aurantifolia* c. Mexican
limon (L.) Burm. f.	FEMMINELLO OLIVA LEMON	*limon* (L.) Burm. f.
aurantifolia (Christm.) Swing.	FLORIDA COMMON LIME	*aurantifolia* (Christm.) Swing.
limetta Risso	FLORIDA SWEET LIME	*limon* (L.) Burm. f.
paradisi Macf.	FOSTER GRAPEFRUIT	*paradisi* Macf.
paradisi Macf.	GRAPEFRUIT	*paradisi* Macf.
paradisi Macf.	HALL GRAPEFRUIT	*paradisi* Macf.
limon (L.) Burm. f.	HARVEY LEMON	*limon* (L.) Burm. f.
sinensis Osbeck	JAFFA ORANGE	*sinensis* (L.) Osbeck
aurantifolia (Christm.) Swing.	KAGHZI LIME	*aurantifolia* (Christm.) Swing.
aurantifolia (Christm.) Swing.	KEY LIME	*aurantifolia* (Christm.) Swing.
nobilis Lour.	KING	*sinensis* X *reticulata*
limon (L.) Burm. f.	LEMON	*limon* (L.) Burm. f.
?	LEONARDY GRAPEFRUIT	? *paradisi* hybrid
aurantifolia (Christm.) Swing.	LIME	*aurantifolia* (Christm.) Swing.
?	LIMEQUAT	*Fortunella* sp. X *C. aurantifolia*
limon (L.) Burm. f.	LISBON LEMON	*limon* (L.) Burm. f.
reticulata Blanco	MANDARIN	*reticulata* Blanco
paradisi Macf.	MARSH SEEDLESS GRAPEFRUIT	*paradisi* Macf.
sinensis Osbeck	MEDITERRANEAN SWEET ORANGE	*sinensis* (L.) Osbeck
aurantifolia (Christm.) Swing.	MEXICAN LIME	*aurantifolia* (Christm.) Swing.
meyerii Y. Tan.	MEYER LEMON	*limon* X *sinensis*
?	MILAM LEMON	?
?	MINNEOLA TANGELO	*paradisi* X *reticulata*
?	MORTON CITRANGE	*P. trifoliata* X *C. sinensis*
?	MURCOTT	?
oblonga Hort. ex Y. Tan.	NAVEL ORANGE	*sinensis* (L.) Osbeck
?	ORLANDO TANGELO	*C. paradisi* X *reticulata*
?	ORTANIQUE	?
?	PAGE	*reticulata* hybrid
?	PALESTINE SWEET LIME	?
?	PERRINE LEMON	?
latifolia Tan.	PERSIAN LIME	hybrid

Tanaka System	Common Name	Swingle System
limonia Osbeck	RANGPUR LIME	*reticulata* var. *austera* hybrid
sinensis Osbeck	RIDGE PINEAPPLE SWEET ORANGE	*sinensis* (L.) Osbeck
?	ROBINSON TANGERINE	*reticulata* hybrid
jambhiri Lush.	ROUGH LEMON	*limon* (L.) Burm. f.
?	ROYAL GRAPEFRUIT	?
?	RUSK CITRANGE	*sinensis* X *P. trifoliata*
?	SAMPSON TANGELO	*paradisi* X *reticulata*
reticulata Blanco	SATSUMA MANDARIN	*reticulata* Blanco
aurantium L.	SEVILLE	*aurantium* L.
grandis (L.) Osbeck	SHADDOCK	*grandis* (L.) Osbeck
limon (L.) Burm. f.	SMOOTH LEMON	*limon* (L.) Burm. f.
aurantium L.	SOUR ORANGE	*aurantium* L.
limetta Risso	SWEET LEMON	*limon* (L.) Burm. f.
limettioides Tan.	SWEET LIME	*limon* (L.) Burm. f.
sinensis Osbeck	SWEET ORANGE	*sinensis* (L.) Osbeck
?	SZIWUIKOM MANDARIN	?
latifolia Tan.	TAHITI LIME	hybrid
?	TANGELO	*reticulata* X *paradisi*
tangerina Hort. ex Tan.	TANGERINE	*reticulata* Blanco
?	TANGOR	*reticulata* X *sinensis*
temple Hort. ex Y. Tan.	TEMPLE	*reticulata* hybrid
?	THORNTON TANGELO	*reticulata* X *paradisi*
?	TRIUMPH GRAPEFRUIT	?
?	TROYER CITRANGE	*sinensis* X *P. trifoliata*
sinensis Osbeck	VALENCIA SWEET ORANGE	*sinensis* (L.) Osbeck
limon (L.) Burm. f.	VILLAFRANCA LEMON	*limon* (L.) Burm. f.
?	WEBBER TANGELO	*reticulata* X *paradisi*
aurantifolia (Christm.) Swing.	WEST INDIAN LIME	*aurantifolia* (Christm.) Swing.
Fortunella Swingle		*Fortunella* Swingle
F. sp.	KUMQUAT	*F.* sp.
japonica (Thumb.) Swing.	MARUMI KUMQUAT	*japonica* (Thumb.) Swing.
?	MEIWA KUMQUAT	? *Citrus* X *Fortunella* hybrid
margarita (Lour.) Swing.	NAGAMI KUMQUAT	*margarita* (Lour.) Swing.
Poncirus Raf.		*Poncirus* Raf.
trifoliata (L.) Raf.	TRIFOLIATE ORANGE	*trifoliata* (L.) Raf.
Severinia Tenore	BOX ORANGE	*Severinia* Tenore
buxifolia (Poir.) Tenore		*buxifolia* (Poir.) Tenore

APPENDIX 3

CONSPICUOUS SYMPTOMS ON PARTS OF TREES AFFECTED BY DISEASES AND DISORDERS OF CITRUS IN FLORIDA

The text's alphabetical arrangement according to common names of diseases is a literary convenience. It offers no assistance in diagnosing troubles in the field or in locating appropriate discussions in the text. The following table is intended to overcome this deficiency. Admittedly, its use does not lead directly to the section concerned since many diseases affect the same part of a tree. Notwithstanding, it is a more realistic approach to identification than a key that, by eliminating all other diseases, leads to a single choice. Keys are inappropriate for separating diseases since symptoms vary with variety, pathogenicity, cross infection, culture, and the environment. Until each disease can be characterized by a single specific test, the best that can be done is to compare the syndromes of several diseases having similar symptoms.

The parts affected are those in which the grower would observe abnormalities. They are not necessarily the parts infected. Thus, for example, the causal organism of mushroom root rot is confined to the roots and crown but secondary symptoms may be found in the tops. Cytologic, histologic, and physiologic abnormalities also occur throughout the tree because of deranged metabolism, but the concern here is with those symptoms that are detectable in the orchard or packinghouse.

Disease or Disorder	Causal Agent	Fruit	Leaves	Twigs & Branches	Trunks	Roots	Tree as a Whole	Remarks
Algal Disease	*Cephaleuros virescens*	Superficial black spotting	Chlorosis, shedding	Bark splitting, dieback			Dieback	
Alternaria Leafspot	*Alternaria citri*		Large brown spots, dropping	Excessive twigging and branching				Primarily a disease in the nursery.
Anthracnose (fruit blemishing only)	*Colletotrichum gloeosporioides*	Spotting						For Anthracnose affecting limes, see Lime Anthracnose. For leaf-spotting of rough lemon and Rangpur lime seedlings, see Alternaria Leafspot.
Arsenic Toxicity	Excess arsenic	Dwarfing, deformation, hardening, internal gumming	Mottling, chlorosis, dropping					Symptoms resemble those of boron deficiency.
Aspergillus Rot	*Aspergillus flavus, A. alliaceus, A. niger*	Spotting, soft rotting						Primarily a postharvest rot.
Biuret Toxicity	Excess biuret		Blotching, chlorosis, marginal necrosis					Cf. **Boron Toxicity** and **Perchlorate Toxicity**.
Black Rot	*Alternaria citri*	Premature coloration, stylar-end spotting, internal rotting, dropping						
Blight	?	Crop reduction	Dulling, rolling, wilting, dropping	Twig and branch dieback, excess sprouting		Eventual rotting	Decline	Usually does not affect trees less than 12 yrs. old. Trees never recover. Cf. Sandhill Decline, Young-Tree Decline.
Blue Mold	*Penicillium italicum*	Soft rot with powdery blue coating						Primarily a postharvest rot. Cf. Green Mold.
Boron Toxicity	Excess boron	Dwarfing, lumpiness, hardening, internal gumming	Tip and marginal chlorosis and necrosis, gumming on under-surfaces, size reduction, dropping					Cf. Fluoride Toxicity.
Brevipalpus Gall	*Brevipalpus phoenicis*		Prevention of refoliation	Galling at nodes of seedlings in nursery				Prolonged infestations lead to death of seedlings.

Disease or Disorder	Causal Agent	Fruit	Leaves	Twigs & Branches	Trunks	Roots	Tree as a Whole	Remarks
Brown Rot	Phytophthora citrophthora, P. parasitica	Leathery, olive-brown rot; drop; pungent odor	Occasional spotting and rotting	Twig dieback				For trunk lesions see Foot Rot.
Cancroid Spot	? Genetic	Canker-like lesions	Pinhead-sized glassy spots, drop	Defoliation, dieback			Decline	Rare. Easy to confuse with Canker.
Canker	Xanthomonas citri	Crater-like pits filled with tan, spongy tissue	Spotting, as under Fruit	Twig spotting, as under Fruit			Decline	Finally eradicated from the USA in 1926.
Cassytha Infestation	Cassytha filiformis	Reduction in yield		Sites of infestation			Decline and stunting after prolonged infestation	Cf. Dodder.
Chilling Injury	Low-temperature storage	Rind collapse, tan to dark brown discoloration						A postharvest problem. Cf. Oleocellosis, Oil-Gland Darkening.
Chimeric Breakdown	? Genetic	Sectorial-chimera-like bands from stem to stylar end, internal gumming						So far seen only in Tahiti (Persian) lime.
Cladosporium Leafspot	Cladosporium oxysporum		Tan to brown spotting, shot-holing					Rare.
Cold Damage	Low temperatures	White crystals on segment walls, oleocellosis-like spotting of rind, drying-out of flesh	Water-soaking, flabbiness, curling, parching, dropping	Dieback of twigs and branches; cold cankers on branches	Cold cankers		Dieback if cold has been severe	Symptoms vary with temperature and duration.
Copper Toxicity	Excess copper	Reduction in yield	Iron deficiency patterns	Dieback		Darkening, stubbiness, scarcity	Growth retarded, decline	Best confirmed by soil analysis for excess copper.
Creasing		Separation of rind along lines of grooving						Cf. Puffing.
Crinkle-Scurf	? Genetic	Reduction in diameter, yield, and juice content	Twisting of blades, faint central chlorotic area	Large branches encircled with bands of corky pimples	Reduction in diameter; patches of corky pimples		Thinning of canopy	So far seen only in trees of late orange varieties, notably Valencia.

APPENDIX 3—Continued

Disease or Disorder	Causal Agent	Fruit	Leaves	Twigs & Branches	Trunks	Roots	Tree as a Whole	Remarks
Crinkly Leaf	Virus		Crinkling, puckering					Symptoms restricted to a few branches. Virus may be present in symptomless carriers. Probably a variant of Infectious Mottling, which see.
Damping Off	Variously *Pellicularia filamentosa, Sclerotium rolfsii, Phytophthora* spp., *Pythium* spp.		Wilting	Rotten collars on seedling stems		Rotting		A disease of plants in the seedbed.
Diplodia Stem-End Rot	*Diplodia natalensis*	Rind rotting at stylar and stem ends; flesh breakdown						A postharvest decay.
Dodder	*Cuscuta americana, C.* spp.			Sites of infestations				May cause galling of stems in lemon and lime trees. Cf. Cassytha.
Endoxerosis	? Water stress	Gum in core, along segment walls, and in inner albedo						Affects primarily lemons and limes.
Exanthema	Copper deficiency	Gummy staining of rind, gum pockets in albedo	Gummy staining, abnormal enlargement	Gum pockets cause bulging of bark on shoots; brown staining of tips; S-shaped growth; multiple budding and sprouting; at times dieback				
Exocortis	Virus	Reduction in yield		Yellow blotching of twigs in susceptible varieties	Bark scaling in susceptible rootstock varieties		Stunting, dieback, decline	Virus may be present in symptomless carriers. Cf. Laminate Shelling.
False Leprosis	?	Leprosis-like lesions	Leprosis-like lesions					Found only on Valencia orange, a variety insusceptible to leprosis. Cf. Leprosis.
Felt	*Septobasidium pseudopedicilatum, S. lepidosaphes*		Superficial felty patches	Superficial felty patches				Nonpathogenic.
Firing	? Water deficiency		Wilting, killing	Partial drying of twigs				Differs from some other wilts in that dead

Disease or Disorder	Causal Agent	Fruit	Leaves	Twigs & Branches	Trunks	Roots	Tree as a Whole	Remarks
								leaves do not drop, affected twigs remain green, and affected parts of canopy are those exposed to prevailing winds.
Fluoride Toxicity	Fluorine gases	Size and yield reduction	Chlorosis, necrosis				Stunting	Cf. Boron Toxicity.
Flyspeck	*Leptothyrium pomi*	Rind blemishing by patches of minute black specks						Cf. Sooty Blotch.
Foot Rot	*Phytophthora parasitica*	For occasional attacks on fruit, see Brown Rot	For occasional attacks on leaves, see Brown Rot		Rotting of bark, usually commencing above bud union and involving more the susceptible scion than the rootstock	Decay of fibrous roots	Decline	Cf. Psorosis, Rio Grande Gummosis, Shell Bark.
Fovea	Virus	Yield reduction	Shedding	Inverse pitting under bark of branches	Inverse pitting under bark of scion		Decline, death	So far observed only in the variety Murcott. Virus may be present in symptomless carriers. Cf. Xyloporosis.
Gas Burn	Ethylene gas and other factors	Brown pitting, sunken areas, ring blemishes						A postharvest problem.
Granulation	?	Drying out of flesh commencing at stem end						Cf. Cold Damage.
Greasy Spot	*Mycosphaerella citri*	For rind blemish, see Pink Pitting	Abaxial edema and tar-like spotting; adaxial chlorosis, necrosis; shedding				Partial defoliation	Cf. Pink Pitting.
Green Mold	*Penicillium digitatum*	Soft rot with green powdery coating; watery halo						Primarily a postharvest rot. Cf. Blue Mold.
Hail Damage	Hailstones	Oil-gland rupturing and oil-burning of rind; pockmarks, rind rupturing, fruit drop	Defoliation	Bark rupturing				

Disease or Disorder	Causal Agent	Fruit	Leaves	Twigs & Branches	Trunks	Roots	Tree as a Whole	Remarks
Infectious Variegation	Virus		Mottling, crinkling, dwarfing					Symptoms restricted to a few branches. Virus may be present in symptomless carriers. Probably a variant of Crinkly Leaf (which see).
Juvenile Spot	?		Gummy platelets on skyward surface; shedding	Twig defoliation			Partial defoliation	Affects trees less than 6 yrs. old.
Laminate Shelling	? Genetic				Laminated corky pimples in a banded pattern on bark of trifoliate orange			Cf. Exocortis.
Leprosis	*Brevipalpus californicus*	Spotting resembling rusty nailheads; dropping	Predominantly marginal gummy platelets; shedding	Scurfy, canker-like lesions on twigs; scaly cambers on branches	Large swollen lesions with thick resinous scales		Decline	Cf. False Leprosis.
Lichens	Various species of lichens		Small superficial patches of various colors	Large superficial patches of various colors	Large superficial patches of various colors			Nonpathogenic.
Lightning Damage	Lightning discharges		Wilting	Twig dieback; killing of bark on twigs between nodes but not at bases of petioles	Killing of narrow strip of bark if tree is hit directly	Destruction of crown if tree is hit directly	Decline; severity of symptoms diminishes with distance of trees from center of strike	
Lime Anthracnose	*Gloeosporium limetticola*	Young fruits and buds blackened and mummified; drop; scabby spots (in older fruits); distortion	Marginal necrosis; distortion	Twig dieback			Decline	**Affects only Key lime types.**
Lime Blotch	? Genetic	Olive bands of sectored rind from stem to stylar end	Chimera-like yellow blotches from midrib to margin	Sunken areas beneath bark	Sunken areas beneath bark		Decline, death	Affects only Tahiti lime and some clones of lemon.
Lumpy Rind	Boron deficiency	Bumps on rind over pockets of gum in albedo						Cf. Arsenic Toxicity.

Disease or Disorder	Causal Agent	Fruit	Leaves	Twigs & Branches	Trunks	Roots	Tree as a Whole	Remarks
Measles	?		Profuse pinhead-sized, brownish spots on undersides of leaves; defoliation					
Melanose	Phomopsis citri	Brown to black pinhead-sized pustules often in a tearstained pattern	Brown to black pinhead-sized pustules imparting sandpapery texture; malformation and drop when heavily attacked	Same pustules on twigs as on fruit and leaves; twig dieback when severely attacked				Cf. Russet.
Mesophyll Collapse	?		Angular translucent chlorosis from midrib to margin, turning to parched areas with age					
Mushroom Root Rot	Clitocybe tabescens	Reduced yield	Sparse	Dieback of twigs and branches	Bark rotting from crown to short distance above groundline	Rotten bark beneath which can be found white, fan-shaped fungal wefts	Decline	Clitocybe mushrooms often appear in season aboveground over attacked roots and crown.
Oil-Gland Darkening	Low temperature storage	Polka-dotted rind blemish; rind scalding (in severe cases)						A postharvest problem. Cf. Chilling Injury.
Oleocellosis	Mechanical rupturing of oil glands	Prominent oil glands in sunken rind areas						
Perchlorate Toxicity	Excess perchlorates in fertilizer		Chlorosis at tips of older leaves; defoliation (in severe cases)					Cf. Biuret Toxicity.
Phomopsis Stem-End Rot	Phomopsis citri	Pliable, tancolored rot beginning at stem end; involves inner and outer surfaces of rind and sometimes the flesh						A postharvest rot. Cf. Diplodia Stem-End Rot.

APPENDIX 3—Continued

Disease or Disorder	Causal Agent	Fruit	Leaves	Twigs & Branches	Trunks	Roots	Tree as a Whole	Remarks
Pink Pitting	Mycosphaerella citri	Minute pink to dark brown necrotic pits in areas between oil glands	The same causal agent produces Greasy Spot (which see)					Conspicuous in grapefruits.
Podagra	Virus	Yield reduction	Sparse	Dieback	Swelling, gumming, and scaling of rough-lemon rootstocks when tops are kumquat		Stunting, decline, death	Symptoms suggest trees are on exocortis-affected trifoliate-orange rootstocks.
Psorosis	Virus							Virus may be present in symptomless carriers. Cf. Foot Rot, Rio Grande Gummosis, Shell Bark.
Type A		Yield reduction	Vein banding and "oakleaf" chlorosis in young leaves	Bark scaling of branches	Bark scaling usually in circular areas		Decline	The common type of psorosis in Florida.
Type B		Partial or complete rings sunken into the rind	As in Type A; also a chlorotic blotching of mature leaves	Bark scaling of branches	Bark scaling usually in elongate areas		Decline	Rarely seen in Florida.
Blind Pocket		Occasional yield reduction	As in Type A		Narrow spindle-shaped depressions paralleling long axis of trunk		Occasional decline	
Concave Gum		Occasional yield reduction	As in Type A	Branch distortion	Pockets longer, wider, and shallower than in Blind Pocket		Occasional stunting	
Puffing	?	Irregular thickening of parts of rind						Cf. Creasing.
Rio Grande Gummosis	?	Yield reduction			Bulging, copious gumming, and scaling of outer layers of bark; gum pockets in wood; affected wood salmon-colored		Decline	Cf. Foot Rot, Psorosis, Shell Bark.

Disease or Disorder	Causal Agent	Fruit	Leaves	Twigs & Branches	Trunks	Roots	Tree as a Whole	Remarks
Robinson Dieback	? *Diplodia natalensis*	Yield reduction	Wilting	Gumming, drooping, and dieback of twigs; occasional invasion of branches	Gumming (in young trees)		Decline	Possibly related to water stress.
Rumple	?	Worm-like settling and discoloration of rind; flesh not affected						A malady of lemons commencing at time of colorbreak.
Russet	*Phyllocoptruta oleivora*	Flat, diffuse, finely grained, brownish staining of the rind	High populations of the causal mite may cause Firing (which see)	Sprouts stained, as under Fruit				Cf. Melanose, Pink Pitting, Greasy Spot.
Salt Burn	Salt toxicity	Drop (in severe cases)	Chronic: Dulling of green color; bronzing. Acute: Tip and marginal chlorosis, necrosis, shedding	In severe cases, twig dieback			Decline (in severe cases)	
Sandhill Decline	?	Small proportion of golf-ball-sized fruit containing aborted seeds and curved cores	Zinc chlorosis speckled with green islets; tip leaves erect, elongated, chlorotic, thickened	Twig dieback			Decline	Cf. Blight, Young-Tree Decline.
Scab	*Elsinoë fawcetti*	Corky, buff to olive-drab eruptions, usually solitary but at times aggregated into patches; when heavily infected, fruits are distorted and may drop	Conical outgrowths with pits at corresponding positions on other side of leaf; distortion and occasional drop	Similar eruptions on tender twigs and pedicels				
Sclerotium Rot	*Sclerotium rolfsii*	A rot of fruit touching the ground						Uncommon. Cf. Brown Rot.
Shell Bark	?	? Reduction in cropping	? Sparse	Scaling and gumming of outer bark of large branches	Scaling and gumming of outer bark, initially above bud union and progressing upward		Decline	Scaling usually begins after lemon trees are more than 7 yrs. old. Cf. Rio Grande Gummosis, Psorosis, Foot Rot.

APPENDIX 3—*Continued*

Disease or Disorder	Causal Agent	Fruit	Leaves	Twigs & Branches	Trunks	Roots	Tree as a Whole	Remarks
Sloughing	?	Paling of rind followed by a rapid, brown rot causing rind to slip						A postharvest disease of grapefruits.
Slow Decline	*Tylenchulus semipenetrans*	Sizes and yields reduced	Dull green to bronze; early wilting during droughts; erect growth; cupping	New twig growth lacks vigor		Parasitized rootlets coated with sand; lack normal white to yellow color	Decline	Cf. Spreading Decline, Slump.
Sooty Blotch	*Gloeodes pomigena*	Rind blemish of dark, cloudy, circular colonies up to ½ in. diameter						Cf. Sooty Mold, Flyspeck.
Sooty Mold	*Capnodium citri, C. citricola*	Rind covered by detachable black film	Leaves covered by detachable black film					Outbreaks proportional to populations of honeydew-secreting insects. Cf. Sooty Blotch.
Sour Rot	*Geotrichum candidum*	Mushy, sour-smelling rot covered with thin, creamy layer of fungal growth						A postharvest rot.
Spanish Moss	*Tillandsia usneoides*			Epiphytic attachment of long trailing masses of grayish-green vegetation				Nonparasitic but objectionable because of its shading effects.
Sphaeropsis Knot	*Sphaeropsis tumefaciens*			Galls at or between nodes, ⅜–3 in. diameter; at first covered with light bark which later becomes dark and cracked; eventually galls are devoid of bark, deeply furrowed, and black	As under Branches		Decline, death	Rare. Most knots in Florida citrus are the result of genetic aberrations.
Spreading Decline	*Radopholus similis*	Reduction in size and number	Sparse, small, pallid	Twig dieback		Scarcity of fibrous roots at depths below 10 in.; lesions on external surfaces of fibrous roots	Decline	Sharp margin at front between affected and nonaffected portions of the orchard. Cf. Slow Decline.

Disease or Disorder	Causal Agent	Fruit	Leaves	Twigs & Branches	Trunks	Roots	Tree as a Whole	Remarks
Star Melanose	Copper sprays on top of melanose pustules	Intensification of melanose blemishes	Intensification and stellate cracking of melanose pustules					
Stem-End Rind Breakdown	Excessive dehydration	Drying and darkening of rind commencing in a ring near the button						A noninfectious post-harvest rind collapse.
Stylar-End Breakdown	? Moisture	At first, a water-soaked but firm rind breakdown at blossom end; later, dehydration, darkening, and settling of affected rind						Found in limes, lime-quats, and lemons.
Stylar-End Russet	?	Superficial blemish of fine corky lines at blossom end						Cf. Russet.
Sunscald	Sunburning	Mild: rind pitting. Severe: hard, gray-black scabs		Scalding, cracking, and sloughing of bark on branches	As under Branches		Decline	May result from spray damage, hedging, or varietal susceptibility.
Tar Spot	Cercospora gigantea	Greasy-spot-like blotches on rind	Greasy-spot-like lesions but with reddish ring inside spots	Lesions on twigs, as under Leaves				Uncommon.
Tatter Leaf-Citrange Stunt	Virus complex		Mottling and distortion		Vertical corrugations in stock portion; grooving around bud union		Decline	Symptoms restricted to certain limes, citranges, and trifoliate orange. Seen thus far in Florida only in experimental trees. Virus complex may be present in symptomless carriers.
Thread Blight	Corticium stevensii	Rind blemished by black, string-like rhizomorphs terminating in black sclerotial bodies	Killing and matting together of leaves which may remain attached by rhizomorphic strands	Rhizomorphic strands on twigs, as under Fruit and Leaves				Uncommon.

APPENDIX 3—Continued

Disease or Disorder	Causal Agent	Fruit	Leaves	Twigs & Branches	Trunks	Roots	Tree as a Whole	Remarks
Tristeza	Virus	Reduction in size and yield	Bronzing; various deficiency patterns; wilting; shedding; vein clearing and cupping in young leaves of limes and certain other species	Dieback	Honeycombing in inner face of bark in susceptible varieties; overgrowth of scion above bud union; stem-pitting in susceptible varieties	Dieback of fibrous and lateral roots	Decline	Generally restricted to trees on sour orange and grapefruit stocks. See diagnostic aids under Tristeza in text. Virus may be present in symptomless carriers.
Wart	?			Branches affected, as under Trunks	Eruptive brown galls to ½ in. diameter and extending cone-shaped deep into the woody cylinder			Rare. Seen only in Valencia trees.
Water Damage	Extreme fluctuations of water table and at times presence of sulfides	Crop reduction	Sparse; small; chlorotic as in N deficiency; when acute, leaves wilt suddenly and drop			Rot, broom-shaped root system	Stunting, decline, death	
Wind Scar	Wind, blowing sand	Rind scarring in various patterns						Copper sprays intensify. Cf. Star Melanose.
Wood Rot Heartwood	Daldinia concentrica, Xylaria polymorpha, Ganoderma sessilis, et al.			Branches affected, as under Trunks	Hollowing-out of woody cylinder, eventuating in wind breakage			
Sapwood	Fomes applanatus	Reduction in yield	Chlorosis, wilting		As above plus attack of sapwood and eruption to bark surface of necrotic, concentrically ringed lesions		Decline	
Xyloporosis	Virus	Reduction in size and yield	Various deficiency patterns; dwarfing; stiff, erect growth	Dieback	Deltoid pitting of woody cylinder and infusion of gum in cambial region and cortex near bud union of susceptible varieties; occasional tilting of trunk; bark	Rot	Stunting, decline	Compare pitting in Tristeza and Fovea. Virus may be present in symptomless carriers. Symptoms appear only in certain mandarins, mandarin hybrids, sweet lime, and lime hybrids. Varieties most commonly affected in Florida are Orlando

Disease or Disorder	Causal Agent	Fruit	Leaves	Twigs & Branches	Trunks	Roots	Tree as a Whole	Remarks
					cracking near bud union in susceptible varieties; overgrowth at bud union when stock is susceptible			and certain other tangelos and sweet lime.
Yellow Spot	Molybdenum deficiency		Elongate yellow spots to ½ in. long, interveinally located, and ascending both sides of midrib in a ladderlike progression; shedding				Decline if persistent	Infrequently seen nowadays.
Yellow Vein	Nitrogen unavailability esp. during winter		Chlorosis involving midribs to entire leaf blades					Usually self-correcting on warming-up of soil and resumption of rains or irrigation.
Young-Tree Decline	?	Small proportion of fruit reduced to size of golf balls and containing aborted seeds and curved cores	Dulling, wilting, delayed flushing, zinc deficiency patterns with green islets; affected branches give rise to dwarfed, erect, leathery, strap-shaped leaves	Dieback			Decline	Cf. Blight, Sandhill Decline.
Zebra-Skin	?	Darkening of rind corresponding with underlying segments; objectionable flavor; prone to early decay						A postharvest problem.

GLOSSARY

Abaxial: Side of the blade away from the axis to which a leaf is attached.

Abscission: The process by which plant parts such as leaves, petals, and fruits are shed by the breakdown of a special layer of cells (the abscission layer).

Acervulus: In certain fungi an asexual **fruiting body** composed of a cushion of fungal strands (**hyphae**) supporting stalks (**conidiophores**) that bear spores (**conidia**); initially formed within host tissue but later breaking through to the surface.

Adaxial: Side of the leaf blade toward the axis to which a leaf is attached.

Albedo: In citrus fruits, the spongy white tissue between the outer colored portion of the rind (the **flavedo**) and the flesh.

Algae: A group of primitive one-celled or multicellular chlorophyll-bearing plants.

Amino acid: One of certain organic compounds containing the amino (NH_2) group and the carboxyl (COOH) group; amino acids are the building blocks of proteins.

Angiosperm: A flowering plant.

Annulus: A ring-like structure in certain mushrooms left near the top of the stalk (**stipe**) after expansion of the cap (**pileus**).

Anthracnose: A plant disease characterized by small spots of dead tissue; a term usually restricted to such spots as are caused by fungi of the Melanconiales.

Apical: At or near the growing tip of a shoot or root.

Apomictic seedling: A seedling arising from a cell other than the egg cell, often in a manner mimicking sexual reproduction; in citrus, such seedlings arise from cells of the nucellus, whence the term **nucellar seedlings.**

Arthropod: Any member of the phylum Arthropoda which includes insects, mites, and certain other invertebrate organisms.

Ascigerous stage: That portion of the life cycle of an **ascomycetous fungus** which gives rise to sexual spores (**ascospores**).

Ascocarp: The fruiting body of an **ascomycetous fungus** in which sexual spores (**ascospores**) are borne.

Ascomycetous fungi: That group of fungi characterized by the production of sexual spores (**ascospores**) in a sac (**ascus**) within a **fruiting body (ascocarp).**

Ascospore: The sexual spore of an **ascomycetous fungus.**

Ascus: Among **ascomycetous fungi**, the sac containing **ascospores.**

Basidiomycetous fungi: That group of fungi characterized by the production of sexual spores (**basidiospores**) on structures known as basidia.

Basidiospore: The sexual spore of a **basidiomycetous fungus.**

Biciliate: Possessed (as in the case of certain **zoospores**) of two cilia or thread-like structures which function in locomotion.

Biotype: A subdivision of a species or subspecies distinguished by some special physiological (as opposed to **morphological**) character.

Blossom end: A colloquialism for that end of the fruit opposite its end of attachment to the tree; synonymous with **stylar end.**

Bromeliaceous: Relating to the Bromeliaceae, a family of **vascular** plants embracing such members as Spanish moss, pineapple, and various of the so-called "air plants."

Buckhorning: A colloquialism for the cutting back of a tree so that only the trunk and stubs of main branches remain; synonymous with **hatracking.**

Bud union: Juncture of the scion and rootstock portions of the trunk.

Button end: A colloquialism for that pole of the fruit attached to the tree; synonymous with **calyx** end.

Callus: Regenerated tissue that develops over or around a wound.

Calyx: Outermost whorl of a flower; in citrus fruits, the hardened lobes make up the so-called button.

Camber: A swelling or arching above the normal contour (e.g., the bulging of limbs and trunks at old lesions of leprosis).

Cambium: A layer of cells in **vascular** plants that gives rise to wood and bark.

Canker: A term of two meanings: (1) any sharply delimited bark **lesion** usually accompanied by malformation (e.g., cold canker, concentric canker); (2) a specific disease of citrus caused by the bacterium *Xanthomonas citri.*

Canopy: The umbrella-like outer mass of foliage.

Capsule: A type of fruit containing two or more seeds and splitting open after drying.

Chimera: Portion of fruit or bark containing patches of tissue different in genetic constitution from the norm.

Chlorosis: An abnormal paling or blanching of the green color in plants; it may be produced by various agencies such as low light intensities, mineral deficiencies, injuries, and infectious diseases.

Clone: A group of individuals (e.g., trees) generated from a common ancestor by vegetative propagation.

Columella: The central pithy axis of a citrus fruit.

Conidiophore: A specialized filament (**hypha**) in fungi on which is borne asexual spores (**conidia**).

Conidium: An asexual spore of a fungus, arising from a specialized filament (**hypha**) known as a **conidiophore.**

Cortex: The tissues in stems and roots lying between the epidermis and the **vascular** tissues.

Cross protection: The suppression of symptoms of one **virus** by the presence of another; at times this test provides presumptive evidence that the two viruses are related.

Cuticle: In plants and insects, the noncellular layer of waxy material covering the epidermis.

Cytoplasm: The protoplasmic material outside the **nucleus** of a cell.

Degreening: A process utilizing ethylene gas to hasten the color change in citrus fruit; synonymous with **gassing.**

Deuteromycetous fungi: A group of miscellaneous fungi having in common the absence of a sexual stage.

Dorsal: Pertaining to the back or upper surface of an organ or organism.

Drupe: A type of fleshy fruit in which seeds are surrounded by a hard stony layer and the entire fruit is covered by a membranous skin.

Edema: A swelling in plant tissue caused by an excessive accumulation of water as the result of certain environmental factors, infection by organisms, or the injection of toxins by insects and mites.

Entomogenous fungi: Fungi that feed in or upon living insects and mites; synonymous with the colloquial "friendly fungi."

Enzyme: A protein substance formed by a living cell and functioning as a catalyst in metabolic reactions.

Epiphyte: A plant (e.g., Spanish moss) that grows on the surface of another plant but not as a **parasite.**

Epiphytotic: A sudden, widespread outbreak of disease among plants; an epidemic, as it were, that affects plants.

Etiology: The study of the cause of a disease.

Feeder root: One of the rootlets making up the fibrous root system of citrus trees.

Fixed coppers: Variously formulated copper-containing fungicides in which elemental copper has been safened, thus rendering it less **phytotoxic** and easier to use than the copper in Bordeaux mixture; synonymous with "proprietary" and "neutral" coppers.

Flatwoods: Those areas of Florida characterized by low-lying, poorly drained soils.

Flavedo: The outer layers of the rind in citrus fruits that contain the characteristic green, orange, or yellow pigments.

Frenching: The chlorotic leaf pattern resulting from a deficiency of zinc.

Friendly fungi: See **Entomogenous fungi.**

Fruiting body: A general term for the organ that produces the spores of a fungus, **alga,** or **lichen.**

Gassing: A process utilizing ethylene gas to hasten the color change in citrus fruit; synonymous with **degreening.**

Genus: A group of closely related species.

Germ tube: The initial thread (**hypha**) produced by a germinating spore.

Hammock: A ridge of wooded land elevated above the level of the adjacent plain or marsh; a variant of hummuck.

Hardpan: An underlying stratum of hard soil or clay that interferes with drainage.

Hatracking: A colloquialism for the cutting-back of a tree so that only the trunk and stubs of main branches remain; synonymous with **buckhorning.**

Heartwood: The dead, inner **xylem** of a woody stem, trunk, or root.

Hematochrome: The orange coloring matter found in certain **algae.**

Herbicide: A chemical for destroying weeds.

Hesperidin: A glucoside in certain citrus fruits that at freezing temperatures crystallizes out into white masses, conspicuous along segment walls.

Histopathology: The study of disease reactions in tissues of a host or **suscept.**

Honeycombing: A fine pitting of the inner face of the bark near the **bud union;** referring usually to one of the symptoms produced in tristeza.

Honeydew: Secretions of certain insects that provide a food source for such fungi as cause sooty mold.

Hyaline: Colorless and transparent or nearly so.

Hypha: One of the filaments forming the mass (**mycelium**) of a fungal colony.

Imperfect stage: That portion of the life cycle of a fungus, **alga,** or **lichen** during which either asexual spores (e.g., **conidia, zoospores, pycnidiospores**) or no spores at all are produced.

Incubation period: The duration between penetration of host tissue by a **pathogen** and the appearance of symptoms.

Indicator plant: A plant used for demonstrating the presence of **viruses** and **mycoplasmas** in budwood or for identifying environmental influences on plant growth.

Inoculum: Portions of a **pathogen** capable of initiating disease in host tissue.

Instar: The form of an insect between successive molts; for example, the first instar is the form between hatching and the first molt.

Interstock: The sandwiched portion of a trunk that results when a budded tree is **topworked** to a new variety; thus, for example, sweet orange is the interstock in a tree on rough-lemon rootstock that has been rebudded to produce a grapefruit top.

Inverse pitting: Depressions in the inner face of the bark (e.g., as in fovea) as opposed to the more frequently encountered pitting (e.g., **stem pitting**) of the **woody cylinder.**

In vitro: In an artificial environment outside the living organism; in contrast to in vivo, within the living organism.

Larva: In nematodes, any life stage between the embryo and the adult; in insects, any stage between the newly hatched individual and the pupal stage in orders with complex **metamorphosis;** in mites, the 6-legged first **instar.**

Lenticel: A small pore in the bark of woody stems, functioning in the exchange of gases to and from interior tissues.

Lesion: A localized spot of diseased tissue whether caused by an injury, a **pathogen,** or a malfunction in **metabolism.**

Lichen: A plant in which a fungus and an **alga** have combined to produce a species that is considered distinct from either of its components. Lichens are unique in that their form and behavior differ markedly from either of the components existing separately.

Mesophyll: The tissues of a leaf between the upper and lower epidermal layers.

Metabolism: The sum total of all chemical activities of a living organism.

Metamorphosis: Change in form during the development of an insect.

Microflora: The microscopic plant life of an area such as soil, bark, or intestines.

Micron: One-thousandth of a millimeter or 0.000039 inch.

Millimicron: One-thousandth of a **micron.**

mm: Abbreviation for millimeter; one-thousandth of a meter, approximately 0.04 inch.

Monoembryonic varieties: Those citrus types that produce no asexual **(nucellar)** seedlings; thus, all seedlings of monoembryonic varieties are usually genetically and morphologically different from the mother parent and are, therefore, not likely to be true to type.

Morphological: Relating to form and structure of plants and animals.

Mycelium: The mass of filaments **(hyphae)** that make up a fungal colony.

Mycoplasmas: Infectious particles intermediate in size between bacteria and **viruses;** they differ from bacteria in not having rigid cell walls (whence their pleomorphic or many-shaped forms), and they differ from **viruses** in not requiring living host cells for multiplication. Mycoplasmas are the smallest known free-living forms of life.

Necrosis: The death of living cells or tissues.

Neutral coppers: See **Fixed coppers.**

Node: A place on a stem where one or more leaves are attached.

Nucellar seedling: One that originates from an unfertilized cell within the nucellus and thus reproduces the mother plant true to type.

Nucleolus: A small, usually round, deeply staining body within the **nucleus** of a cell; composed of protein and ribonucleic acid.

Nucleus: A round body within the **cytoplasm** of a cell; it contains the cell's hereditary material that regulates structure and function.

Nymph: A stage in the life cycle of certain **arthropods** (e.g., 8-legged mites) between the larva and the adult.

Oospore: A sexual spore of a **phycomycetous fungus.**

Parasite: An organism that lives on or in another living organism to obtain its nutritional needs and water requirements. The term "parasite" emphasizes this dependency whereas the term **pathogen** refers to an infectious organism that may cause pathological effects other than those due only to food and water withdrawal.

Parenchymatous: Referring to the parenchyma, a soft tissue made up of thin-walled, living cells.

Parthenogenesis: Reproduction by the development of an unfertilized egg cell.

Pathogen: A living organism (e.g., a fungus) or life-like organism (e.g., a **virus**) capable of causing disease; compare **parasite.**

Pedicel: The stalk of an individual flower and later the stem of a fruit.

Perfect stage: That portion of the life cycle of an **alga,** fungus, or **lichen** during which sexual spores (i.e., those resulting from the fusion of egg and sperm cells or by **parthenogenesis**) are formed. Examples of sexual spores are **ascospores, oospores,** and **basidiospores.**

Pericycle: The ring of cells bordering the **vascular** system of plants; in roots, the tissue immediately inside the endodermis and outside the primary **vascular** tissues.

Periderm: In woody stems, trunks, and roots, the cork, cork cambium, and **phelloderm** collectively.

Perithecium: A flask-shaped or spherical **ascocarp** with an opening; it contains the **ascospores.**

Petiole: The stalk that attaches leaf to stem; in citrus, it includes the wing, that leaf-like structure subtending the blade proper.

pH: A measure of the acidity or alkalinity of a solution in which 7 represents neutrality, 0 maximum acidity, and 14 maximum alkalinity; literally, p(otential of) H(ydrogen).

Phelloderm: The soft, green cortical tissue that forms on the inner side of the phellogen in certain plants.

Phloem: That portion of the **vascular** system involved chiefly in the downward conduction of photosynthesized food; compare **xylem.**

Phycomycetous fungi: So-named from a supposed relationship with the **algae.** In this heterogeneous group, many are either aquatic or favored by damp conditions and produce swimming spores.

Phytopathogenic: Capable of producing disease in plants.

Phytophagous: Feeding on plants; said especially of certain insects, mites, and nematodes.

Phytotoxic: Damaging to plants, said usually of agricultural chemicals.

Pileus: The umbrella-like cap of a stalked mushroom.

Polyphagous: Feeding on or utilizing a variety of plants or food; said usually of insects, mites, and nematodes.

Pustule: A localized swelling of the plant epidermis which may rupture to expose the **pathogen** or by-products of infection.

Pycnidiospore: An asexual spore borne within a flask-shaped fruiting body produced by certain fungi and **lichens.**

Pycnidium: In certain fungi and **lichens,** a globose or flask-shaped fruiting body containing asexual spores **(pycnidiospores).**

Race: A variety of a species that is similar in form to that species but different in other respects.

Rhizomorph: A thread- or cord-like structure composed of aggregated fungal filaments **(hyphae);** compare **xylostroma.**

Rhizosphere: That portion of the soil occupied by living roots.

Ridge: That central portion of peninsular Florida characterized by rolling, sandy hills.

Rosette: A plant disease symptom in which the stem or stem tip is shortened, thereby leading to an abnormal clustering of the leaves.

Rutaceous: Of the Rutaceae, the family to which *Citrus* belongs.

Sandhill locations: A colloquialism for the rolling, well-drained areas of peninsular Florida.

Sandsoak: A colloquialism for an area of soil composed of fine sand particles; usually unfavorable for the growing of citrus.

Saprophagous: Feeding on dead or decaying matter.

Saprophyte: A plant, fungus, or bacterium that lives on dead or decaying organic matter.

Sapwood: The secondary **xylem** in woody stems and trunks that is still active and contains some living cells.

Sclerotium: Among some fungi, a resting body, usually dark-colored, from which **mycelium** or various fruiting structures may develop.

Secondary invader: A weakly- or non-pathogenic organism that becomes established in tissues damaged by **pathogens** or injuries.

Sector: A genetic alteration in leaves or fruits resulting in a form of growth that differs from the normal.

Senescent: Old-aged, overmature.

Sessile: Stalkless; attached directly to a base.

Shotholing: Perforations in the leaf blade caused by the dropping out of necrotic tissue.

Skirt: That portion of a tree's canopy nearest the ground.

Somatic mutation: A genetic alteration affecting body (as opposed to sexual) cells.

Specialty fruit: A colloquialism for citrus varieties other than the standard, widely cultivated round oranges and grapefruit; examples are tangelos, Temples, and Murcotts.

Sporangium: A **fruiting body** containing asexual spores.

Stalked: Borne on a stem; attached to a base by an elongated structure; compare **sessile.**

Stele: The central core of **vascular** tissue in a plant root.

Stem pitting: Small indentations in the face of the **woody cylinder** of trees; though the condition may be caused by various diseases, the term is usually restricted to the pitting produced in tristeza.

Stipe: The stem supporting the cap **(pileus)** of a mushroom.

Stomate: One of the microscopic pores in the epidermal surface of plants through which pass water vapor and gases.

Stylar end: That pole of a fruit opposite the end of attachment; colloquially **blossom end.**

Suscept: A plant susceptible to attack by a **pathogen.** The suscept-pathogen relationship differs from that of the host-parasite relationship in covering all pathologic effects, not just those due to parasitism; thus, for example, citrus is a suscept of the pathogen *Cephaleuros virescens* but parasitism is not involved.

Thallophyte: Any plant of the subkingdom Thallophyta which includes the bacteria, fungi, **algae,** and **lichens.**

Thallus: The undifferentiated leafless, stemless, rootless plant body as exemplified by the **thallophytes.**

Tissue culture: The cultivation under sterile conditions of pieces of tissue or organs on an artificial medium.

Topworking: A colloquialism for the process of rebudding or regrafting the top of a tree to another variety.

Vacuole: Any small cavity in the protoplasm of a cell.

Vascular: Relating to vessels for the transmission of plant fluids.

Vector: A living organism capable of transmitting a **pathogen;** said especially of insects, mites, or nematodes that transmit a **virus** or **mycoplasma** from one plant to another.

Veil: The membrane covering the caps of certain mushrooms; in some species, its rupture leaves a ring **(annulus)** around the stalk **(stipe)** of the cap **(pileus).**

Vein clearing: A narrow, faint chlorotic emargination of the veins in leaves; often indicative of **virus** infection; synonymous with vein banding.

Virus: A nucleoprotein capable of causing infection and multiplying in a host.

Vulva: The external opening of the female reproductive system.

Woody cylinder: The portion of a tree trunk under the bark.

Xylem: The water-conducting and supporting tissues of **vascular** plants.

Xylostroma: The blackish, hardened cord-like strands of compacted **mycelium** that become visible in the cracks of roots infected by mushroom root rot.

Zoospore: A swimming asexual spore as in certain of the **phycomycetous fungi** and the **algae.**